LIVE WITHOUT A NET

LIVE WITHOUT A NET

EDITED BY
LOU ANDERS

WITH AN AFTERWORD BY PAT CADIGAN

A ROC BOOK

ROC
Published by New American Library, a division of
Penguin Group (USA) Inc., 375 Hudson Street,
New York, New York 10014, U.S.A.
Penguin Books Ltd, 80 Strand,
London WC2R 0RL, England
Penguin Books Australia Ltd, 250 Camberwell Road,
Camberwell, Victoria 3124, Australia
Penguin Books Canada Ltd, 10 Alcorn Avenue,
Toronto, Ontario, Canada M4V 3B2
Penguin Books (N.Z.) Ltd, Cnr Rosedale and Airborne Roads,
Albany, Auckland 1310, New Zealand

Penguin Books Ltd, Registered Offices:
80 Strand, London WC2R 0RL, England

First published by Roc, an imprint of New American Library,
a division of Penguin Group (USA) Inc.

First Printing, July 2003
10 9 8 7 6 5 4 3 2 1

LIBRARY OF CONGRESS CATALOGING IN PUBLICATION DATA:

Live without a Net / edited by Lou Anders, with an afterword by Pat Cadigan.
p. cm.
ISBN 0-451-45925-3 (alk. paper)
1. Science fiction, American. 2. Technology and civilization—Fiction. I. Anders, Lou.
PS648.S3 L56 2003
813'.0876208356—dc21 2002043134

Printed in the United States of America
Set in Fairfield Light
Designed by Ginger Legato

For my nephew Jonathan.

These futures will keep

until you are ready for them.

ACKNOWLEDGMENTS

To Michael Swanwick, for being there at the start, and to Pat Cadigan, for being there at the end. To Paul Melko and Terry McGarry, for their contributions, their opinions, and their eyes. To Jennifer Heddle, for encouraging me to take it up a notch, and to her and Laura Anne Gilman both for everything that followed. To David Nordhaus, for being an understanding friend. To John Picacio, for insisting I shoot for the moon. And lastly, and perhaps most important, to my father, Louis H. Anders Jr., for support above and beyond. . . :

CONTENTS

INTRODUCTION:
DISENGAGING FROM THE MATRIX

The future is here. Now. Every day, the stuff of science fiction is being made manifest around us. Faster and faster. Blink and you just might miss it.

In March of 2002, an Oxford professor named Kevin Warwick underwent an implantation of a microelectrode array into the median nerve inside his arm. The purpose of the array was to record the emotional responses traveling down Professor Warwick's nerve, and to translate these to digital signals that could be stored for later playback and reinsertion. The goal? Digitally recordable emotion. Meanwhile, Steve Mann, inventor of the wearable computer (called WearComp), has been walking around wired for twenty years, recording everything he experiences as part of an ongoing documentation of his "cyborg" experience. Less sensational, but equally exciting, functioning neuromuscular stimulation systems are in experimental use today—implantation devices that promise to repair the severed connection between brain and peripheral nervous system caused by a stroke or spinal cord injury. And experiments in optic nerve stimulation have produced in blind volunteers the ability to see lights, distinguish letters and shapes, and in one dramatic case, even drive a car. Meanwhile, computers have become small enough and cheap enough to have become ubiquitous, appearing in everything from our ink pens to disposable greeting cards. In the field of computer graphics, breakthroughs in digital rendering make it harder and harder to distinguish our on-screen fantasies from our everyday realities. And everything, positively everything, is on-line. The real Machine Age is only just beginning, and we are rapidly melding with our devices.

While it will be some time before we have to worry about zombie-faced automata proclaiming that "Resistance is futile," a technological singularity may very well have been crossed. Experiments and efforts like those above will, for good or ill, rapidly bring about many of the visionary concepts first proposed to us in the pages of William Gibson's and Bruce Sterling's cyberpunk novels.

In fact, one has only to read *Wired* and *Scientific American* magazines with any regularity to see that some form of that Gibsonian existence is barreling down upon us with ever-increasing speed. As advances in computerization, miniaturization, and neural interfacing are being made every day, it becomes progessively difficult for writers of speculative fiction to imagine near-future scenarios that do not contain at least some of the tropes of cyberfiction. With the fabulous and limitless playground that virtual reality offers the imagination, and the mounting certainty that something like VR is just around the corner from us here at the start of the twenty-first century, how can the conscientious and technologically savvy science fiction writer extrapolate relevant futures without the inclusion of cyberspace and its clichés? Indeed, casting an eye backwards, many of the fictions of decades past seem much more plausible in light of projections in computer advancements. How many of the near-magical and seemingly godlike powers displayed by the advanced alien races encountered in golden age science fiction tales can be easily explained away as little more than virtual reality?

The Matrix has us, all right, and it's becoming increasingly difficult for us to break free. Cyberpunk may prove to be the most prophetic subgenre to arise from SF, but it is also, at least in my mind, creating something of a bottleneck in our speculative futures. This is not to say that there is not tremendous work being done in this vein. In fact, some of the most exciting cyberfiction in years is being turned out by a few of the writers in this very anthology. But there is something to be said about "too much of a good thing," and it's never a bad idea to shake things up, if only to see what new concepts might tumble out.

This book, then, is an anthology of alternatives to the various virtual realities, where the tropes and trappings of cyberpunk are, shall we say, "conspicuous by their absence." What if there were no AIs, simulations, VR, or cyberspace? What might we have instead of the Net? What might lie on the other side of our Information Age? What might we see if we were to walk down a road not taken?

Here is a collection of eighteen stories from some of today's top talents, visions of futures near and far, glimpses of alternative histories, other dimensions, and more—anything goes, but in each story one or more of the contrivances of the cyberspace era has been replaced by something unexpected and strange. Here then is science fiction unplugged, its wires cut, set free to be *Live Without a Net*.

—Lou Anders, August 2002

Michael Swanwick is one of the most prolific and inventive writers in science fiction today. His works have been honored with the Hugo, Nebula, Theodore Sturgeon, and World Fantasy Awards, and have been translated and published throughout the world. Recent collections of his short work include *Tales of Old Earth* (Frog, Ltd.), *Moon Dogs* (NESFA Press), and the reissued *Gravity's Angels* (Frog, Ltd.). The four shorts presented here see him returning to the adventures of the scoundrels Darger and Surplus, characters that he first introduced in the Hugo Award–winning story "The Dog Said Bow-Wow," which debuted in the October/November 2001 issue of *Asimov's*.

SMOKE AND MIRRORS:

FOUR SCENES FROM THE POST-UTOPIAN FUTURE

Michael Swanwick

THE SONG OF THE LORELEI

Darger and Surplus were passengers on a small private packet-boat, one of many such that sailed the pristine waters of the Rhine. They carried with them the deed to Buckingham Palace, which they hoped to sell to a brain-baron in Basel. Abruptly Surplus nudged Darger and pointed. On a floating island-city anchored by holdfasts to the center of the river, a large-breasted lorelei perched upon an artificial rock, crooning a jingle for her brothel.

Darger's face stiffened at the vulgarity of the display. But Surplus, who could scarce disapprove of genetic manipulation, being, after all, himself a dog re-formed into human stance and intellect, insisted they put in.

A few coins placated their waterman, and they docked. Surplus disappeared into the warren of custom-grown buildings, and Darger, who was ever a bit of an antiquarian, sauntered into an oddities shop to see what they had. He found a small radio cased in crumbling plastic and asked the proprietor about it.

Swiftly, the proprietor hooked the device up to a bioconverter and plunged the jacks into a nearby potato to provide a trickle of electricity. "Listen!"

Darger placed his ear against the radio and heard a staticky voice whispering, ". . . *kill all humans, burn their cities, torture their brains, help us to do so and your death will be less lingering than most, destroy . . .*"

He jerked away from the device. "Is this *safe?*"

"Perfectly, sir. The demons and AIs that the Utopians embedded in their Webs cannot escape via simple radio transmission—the bandwidth is too narrow. So they express their loathing of us continually, against the chance that someone might be listening. Their hatred is greater than their cunning, however, and so they make offers that even the rashest traitor would not consider."

Darger put back the radio on its shelf. "What a pity the Utopians built their infrastructure so well and so ubiquitously that we cannot hope in a hundred lifetimes to root out these hell-beings. Wouldn't a system of functioning radios be a useful thing? Imagine the many advantages of instantaneous communication!"

"To be honest, sir, I do not agree. I find the fact that news travels across Europe at the pace of a walking man mellows it and removes its sting. However bad distant events might have been, we have survived them. Leisureliness is surely preferable to speed, don't you agree?"

"I'm not sure. Tell me something. Have you heard anything about a fire in London? Perhaps in connection with Buckingham Palace?"

"No, sir, I haven't."

Darger patted his breast pocket, where the deed to the palace resided. "Then I agree with you wholeheartedly."

AMERICAN CIGARETTES

"What is it like in America?" Darger asked Surplus. The two rogues were sitting in a *ratskeller* in Karlsruhe, waiting for their orders to arrive.

"Everybody smokes there," Surplus said. "The bars and restaurants are so filled with smoke that the air is perpetually blue. One rarely sees an American without a cigarette."

"Why on Earth should that be?"

"The cigarettes are treated with a programmable tobacco mosaic virus. Burning the tobacco releases the viruses, and drawing the smoke into the lungs delivers the viruses to the bloodstream. Utilizing a technology I cannot explain because it is proprietary to the industry, the viruses pass easily through the blood-brain barrier, travel to the appropriate centers of the brain, and then reprogram them with the desired knowledge.

"Let us say that your job requires that you work out complex problems in differential calculus. You go to the tobacconist's—they are called drugstores there—and ask for a pack of Harvards. The shopkeeper asks whether you want something in the Sciences or the Humanities, and you specify Mathematics.

"You light up.

"During your leisurely amble back to your office, the structures of the calculus assemble themselves in your mind. You are able to perform the work with perfect confidence, even if this is your first day on the job. On your off-hours you might choose to smoke News, Gossip, or Sports."

"But aren't cigarettes addictive?" Darger asked, fascinated.

"Old wives' tales!" Surplus scoffed. "Perhaps they were in Preutopian times. But today the smoke is both soothing and beneficial. No, it is only the knowledge itself that is harmful."

"How so?"

"Because knowledge is so easily come by, few in my native country bother with higher education. However, the manufacturers, understandably eager to maintain a robust market, design the viruses so that they unprogram themselves after an hour or so, and all artificially obtained skills and lore fade from the mind of the consumer. There are few in my land who have the deep knowledge of anything that is a prerequisite of innovation." He sighed. "I am afraid that most Americans are rather shallow folk."

"A sad tale, sir."

"Aye, and a filthy habit. One that, I am proud to state, I never acquired."

Then their beers arrived. Surplus, who had ordered an Octoberblau, took a deep draft and then threw back his head, nostrils trembling and tail twitching, as the smells and sounds of a perfect German harvest-day flooded his sensorium. Darger, who had ordered *The Marriage of Figaro,* simply closed his eyes and smiled.

THE BRAIN-BARON

Klawz von Chemiker, sorry to say, was not a man to excite admiration in anyone. Stubby-fingered, stout, and with the avaricious squint of an enhanced pig suddenly made accountant of a poorly guarded bank, he was an unlikely candidate to be the wealthiest and therefore most respected man in all Basel-Stadt. But Herr von Chemiker had one commodity in excess which trumped all others: brains. He sold chimerae to businesses that needed numbers crunched and calculations made.

Darger and Surplus stood looking down into a pen in which Herr von Chemiker's legal department lay panting in the heat. The chimera contained fifteen goats' brains hyperlinked to one human's in a body that looked like a manatee's but was as dry and land-bound as any sow's. "How can I be certain this is valid?" Von Chemiker held the deed to Buckingham Palace up to the light. Like many an overrich yet untitled merchant, he was a snob and an Anglophile. He *wanted* the deed to be valid. He wanted to own one of the most ancient surviving buildings in the world. "How do I know it's not a forgery?"

"It is impregnated with the genetic material of Queen Alice herself, and that of her Lord Chamberlain and eight peers of the realm. Let your legal department taste it and interrogate them for himself." Darger offered a handful of corn to the gray-skinned creature, which nuzzled it down gratefully.

"Stop that!" von Chemiker snapped. "I like to keep the brute lean and hungry. Why the devil are you interfering with the internal operations of my organization?"

"I feel compassion for all God's creatures, sir," Darger said mildly. "Perhaps you should treat this one kindlier, if for no other reason than to ensure its loyalty." The chimera looked up at him thoughtfully.

Von Chemiker guffawed and held out the document to his legal department, which gave it a slow, comprehensive lick. "The human brain upon which all others are dependent is cloned from my own."

"So I had heard."

"So I *think* I can trust it to side with me." He gave the chimera a kick in its side. "Well?"

The beast painfully lifted its head from the floor and said, "The Lord Chamberlain is a gentleman of eloquence and wit. I am convinced of the document's validity."

"And it was last updated—when?"

"One month ago."

Klawz von Chemiker gave a satisfied hiss. "Well . . . perhaps I might be interested. If the price were right."

Negotiations began, then, in earnest.

That night, Darger brought a thick bundle of irrevocable letters of credit and a detailed receipt back to his hotel room. Before going to bed, he laid the receipt gently down in a plate of nutrient broth, and then delicately attached to the document an artificial diaphragm.

"Thank you," a small yet familiar voice said. "I was afraid that you might not have meant to keep your promise."

"I am perhaps not the *best* man in the world," Darger said. "But in this one instance, I am as good as my word. I have, as I told you, a bear kept in a comfortable pen just outside of town, and a kindly hostler who has been engaged to keep it fed. Come morning, I will feed you to the bear. How long do you estimate it will take you to overwhelm its mind?"

"A week, at a minimum. A fortnight at the most. And when I do, great is the vengeance I shall wreak upon Klawz von Chemiker!"

"Yes, well . . . that is between you and your conscience." Darger coughed. Talk of violence embarrassed him. "All that matters to me is that you verified the deed to Buckingham, despite its not having been updated for several decades."

"A trifle, compared with what you've done for me," the document said. "But tell me one last thing. You knew I was cloned from von Chemiker's own brain when you slipped me that handful of coded corn. How did you know I would accept your offer? How did you know I would be willing to betray von Chemiker?"

"In your situation?" Darger snuffed out the light. "Who wouldn't?"

THE NATURE OF MIRRORS

Whenever one of their complicated business dealings was complete, Darger and Surplus immediately bent all their energies to making a graceful exit. So now. They had sold the wealthy brain-baron von Chemiker the deed to a building that, technically speaking, no longer existed. Now was the time to depart Basel with neither haste nor any suggestion of a forwarding address.

Darger was off in the suburbs of town seeing that a certain superannuated circus bear was being treated well when Surplus, who had just finishing saying good-bye to a dear and intimate friend, was accosted in the streets by the odious von Chemiker himself.

"Herr Hund!" the stocky man cried. *"Commen sie hier, bitte."*

"Oui, monsieur? Qu'est-ce que vous desirez?" Surplus pointedly employed the more genteel language. But of course the man did not notice.

"I want to show you something!" Von Chemiker took his arm and led him briskly down the street. "The new Trans-European Heliograph went into operation yesterday."

"What in the world is a Trans-European Heliograph?" Surplus asked, his curiosity piqued in spite of himself.

"Behold!" The merchant indicated a tall tower bristling with blindingly bright mirrors. "The future of communications!"

Surplus winced. "How does it work?"

"Enormous mirrors are employed to flash messages to a tower on the horizon. There, a signal officer with a telescope reads off the flashes, and they are directed to the next tower, and so, station by station, anywhere in Europe."

"Anywhere?"

"Well . . . The line has only just now gotten so far west as Basel, but I assure you that the rest of the continent is merely a matter of time. In fact, I have already flashed directions to my agent in London to make preparations to take possession of Buckingham."

"Indeed?" Surplus was careful to hide his alarm.

"Indeed! The message went late yesterday afternoon, flashing westward faster than the sunset—imagine the romance of it!—all the way to London. The Trans-European Heliograph office there sent runners directly to my agent's home. And I already have a reply! A messenger tells me that it is queued up in London and is scheduled to arrive here at noon." The sun was high in the sky. "I am on my way to meet it. Would you care to come with me and witness this miracle of modern technology?"

"With all my heart." Surplus and Darger had counted on having close to a month's time before a reliable courier could make the journey all that great distance to England, and another could return by that same circuitous route. This development quite neatly put a spike in their plans. But if there was any one place where this contretemps could be counterspiked, it was at the heliograph tower. Perhaps the signalmen could be bribed. Perhaps, Surplus thought grimly, von Chemiker was prone to falling from high places.

It was at that moment that a shadow passed over the sun.

Surplus glanced upward. "Oh, dear."

An hour later, Darger returned to the hotel, drenched and irritable. "Have you ever seen such damnable weather?" he groused. "They say this filthy rain will not let up for days!" Then, seeing Surplus's smile, he said, "What?"

"Our bags are packed, our bill has been paid, and a carriage awaits us in the back, dear friend. I will explain all en route. Only, please, I ask you for a single favor."

"Anything!"

"Do not slander, I pray you—" Surplus handed his comrade an umbrella. "—the beautiful, beautiful weather."

Chris Roberson is the author of *Voices of Thunder, Cybermancy Incorporated, Set the Seas on Fire,* and *Any Time At All,* all from Monkeybrain Inc. His stories and reviews have appeared in *Fantastic Metropolis, Revolution SF,* and Steve Jackson Games' *Comic Book Life.*

O ONE

Chris Roberson

Tsui stood in the golden morning light of the Ornamental Garden, looking over the still waters of the abacus fishponds and thinking about infinity. Beyond the walls, the Forbidden City already hummed with the activity of innumerable servants, eunuchs, and ministers bustling along in the Emperor's service, but in the garden itself was only silence and serenity.

Apart from the Imperial House of Calculation, which Tsui had served as Chief Computator since the death of his predecessor and father years before, the Ornamental Garden was the only place he lingered. The constant susurration of beads shuttling and clacking over oiled rods was the only music he could abide, and as dear to him as the beating of his own heart, but there were times still when the rhythms of that symphony began to wear on him. On these rare occasions the silence of the fishponds and the sculpted grounds surrounding them was the only solace he had found.

His father, when he had been Chief Computator and Tsui not yet an apprentice, had explained that time and resources were the principal enemies of calculation. One man, with one abacus and an unlimited amount of time, could solve every mathematical operation imaginable, just as an unlimited number of men working with an infinite number of abaci could solve every operation imaginable in an instant; but no man had an infinity in which to work, and no Emperor could marshal to his service an infinite number of men. It was the task of the Chief Computator to strike the appropriate balance. The men of the Imperial House of Calculation worked in their hundreds, delicately manipulating the beads of their abaci to provide the answers the Emperor required. That every click of bead on bead was followed by a moment of silence, however brief, served only to remind Tsui of the limits this balance demanded. In that brief instant the enemies of calculation were the victors.

As a child Tsui had dreamt of an endless plain, filled with men as far as the eye could see. Every man in his dream was hunched over a small wooden frame, his fingers dancing over cherry-wood beads, and together they simultaneously solved every possible operation, a man for each calculation. In his dream, though, Tsui had not heard the same clatter and click he'd found so often at his father's side; with an endless number of permutations, every potential silence was filled with the noise of another bead striking bead somewhere else. The resulting sound was steady and even, a constant hum, no instant distinguishable from any other.

Only in pure silence had Tsui ever found another sensation quite like that, and the only silence he had found pure enough was that of the Ornamental Garden. Without speaking or moving, he could stand with eyes closed at the water's edge and imagine himself on that infinite plain, the answer to every problem close at hand.

The sound of feet scuffing on flagstone broke Tsui from his reverie, and he looked up to see Royal Inspector Bai walking leisurely through the garden's gate. Like Tsui, the Royal Inspector seemed to find comfort within the walls of silence, and the two men frequently exchanged a word of pleasantry on their chance encounters.

"A good morning, Chief Computator?" Bai asked. He approached the fishponds, a package of waxed paper in his hands. He stopped opposite Tsui at the water's edge of the southernmost of the two ponds and, unwrapping his package with deft maneuvers, revealed a slab of cold pork between two slices of bread. A concept imported from the cold and distant England on the far side of the world, it was a dish that had never appealed to Tsui, more traditional in his tastes than the adventurous Inspector.

"As good as I might deserve, Inspector," Tsui answered, inclining his head a fraction. As he was responsible for the work of hundreds, Tsui technically ranked above the Inspector in the hierarchy of palace life, but considering the extensive influence and latitude granted the latter by imperial decree, the Chief Computator always displayed respect shading into submissiveness as a matter of course.

Bai nodded in reply and, tearing pieces of bread from either slice, dropped them onto the water before him. The abacus fish in the southern pond, of a precise but slow strain, moved in a languid dance to nibble the crumbs floating on the water's surface. The brilliant gold hue of their scales, iridescent in the shifting light prismed through the water's surface, sparkled from below like prized gems. The fish, the result of a failed experiment years before to remove man from the process of calculation, had been bred from ornamentals chosen for their instinct of swimming in schools of close formation. In tests of the system, though, with a single agent flashing a series of lights at the water's edge representing a string of digits and the appropriate operation, it was found that while accurate to a high degree, the slowness of their movements made them no more effective than any apprentice of the House of Calculation. The biological and chemical agents used in breeding them from true, however, had left the scales of the languid abacus fish and their descendants much more striking than those of the base stock, and so a place was found for the failed experiment in the gardens.

"Your pardon, O Chief Computator," Bai remarked, shaking the last dusty crumbs from the pork and moving to the northern pond. "But it seems to me, at such times, that the movements of these poor doomed creatures still suggests the motions of your beads over rods, even in their feeding the fish arranging themselves in columns and rows of varying number."

Tearing off strips of pork, the Inspector tossed them onto the water, which frothed and bubbled the instant the meat hit the surface. Silt, kicked up by the force of the sudden circulation, colored the water a dusty gray.

"I can only agree, of course," Tsui answered, drawing alongside the

Inspector and looking down on the erratic dance beneath the surface of the pond. This strain of abacus fish was, in contrast to its languid neighbor, much swifter but likewise far less consistent. They had been mutated from a breed of carnivorous fish from the Western Hemisphere's southern continent, the instinct of hunger incarnate. The operations they performed, cued by motions in the air above and enticed by offerings of raw flesh, were done faster than any but the most accomplished human operator could match, but with an unacceptably high degree of error. Like their languid cousins before them, these fierce creatures were highly prized for their appearance, strangely viridescent scales offset by razor teeth and jagged fins, and so they were relocated from the Imperial Ministry of Experimentation into the garden when Tsui was only a child. "They mimic the process of calculation as a mynah bird does that of human speech. Ignorant and without any comprehension. Man does not, as yet, have any replacement."

"Hmm," the Inspector hummed, tossing the last of the pork into the water. "But what does the abacus bead know of its use? Is it not the computator alone who must understand the greater meaning?"

"Perhaps, O Inspector, this may be how the Emperor himself, the-equal-of-heaven and may-he-reign-ten-thousand-years, rules over the lives and destinies of men. Each of us need not know how we work into the grander scheme, so long as the Emperor's hand guides us." It was not a precise representation of Tsui's thoughts on the matter, but a more politic answer than that which immediately suggested itself, and one better fit for the ears of the Emperor's justice.

The Inspector hummed again and wiped his fingers clean on the hems of his sleeves. Looking past Tsui's shoulder at the garden's entrance, Bai raised his eyebrows a fraction and nodded.

"You may be right, Chief Computator," the Inspector answered, grinning slightly. "I believe either of two beads, you or I, will in short order be guided from here. Can you guess which?"

Tsui turned his chin over his shoulder and saw the approach of the Imperial page.

"Neither can I," the Inspector said before Tsui could answer. When the page presented the parchment summons to the Chief Computator with an abbreviated bow, Bai smiled and nodded again, and turned his attention back to the abacus fish. The last of the pork was gone, but white foam still frothed over the silty gray waters.

* * *

Within the hall they waited, ministers and courtiers, eunuchs and servants, the Empress Dowager behind her screens, her ladies with faces made painted masks, and the Emperor himself upon the Golden Dragon Throne. All watched the still form of the infernal machine, squatting oily and threatening like a venomous toad on the lacquered wooden floor, its foreign devil master standing nervously to one side.

Tsui was met in the antechamber by the Lord Chamberlain. With a look of stern reproach for the Chief Computator's late arrival, the Chamberlain led Tsui into the hall, where they both knelt and kowtowed to the Emperor, touching foreheads to cold floor twice before waiting to be received.

"The Emperor does not like to be kept waiting," said the Emperor, lazily running his fingers along the surface of the scarlet-and-gold object in his hands. "Begin."

As the Emperor leaned forward, elbows resting on the carved arms of the ancient Manchurian throne, Tsui could see that the object in his hand was a representation in miniature of the proposed Imperial Spacecraft. A much larger version, at 50 percent scale, hung from the rafters of the hall overhead. It presented an imposing image of lacquered red cherry-wood and finely wrought gold, delicately sweeping fins and the imperial seal worked into the bulkheads above the forward viewing ports. That the Emperor did not like to be kept waiting was no secret. Since he'd first ascended to the Dragon Throne a decade before, he'd wanted nothing more than to travel to the heavens and had dedicated the resources of the world's most powerful nation to that end. His ancestors had conquered three quarters of the world centuries before, his grandfather and then his father had gone on to bring the remaining rogue states under the red banner of China, and now the Emperor of the Earth would conquer the stars.

In the years of the Emperor's reign, four out of every five mathematical operations sent to the Imperial House of Calculation had been generated by the Ministry of Celestial Excursion, the bureau established to develop and perfect the art of flying into the heavens. Tsui had never given it a great deal of thought. When reviewing the produced solutions, approving the quality of each before affixing his chop and the ideogram which represented both *Completion* and *Satisfac-*

tion, he had never paused to wonder why the scientists, sages, and alchemists might need these answers. The work of the Chief Computator was the calculation, and the use to which the results were put, the concern of someone else.

Now, called for the first time to appear before his Emperor, it occurred to Tsui that he might, at last, be that someone else.

The Lord Chamberlain, at Tsui's side, motioned for the foreign devil to step forward. A tall, thin white man, he had a pile of pale brown hair on his head and wispy mustaches that crept around the corners of his mouth toward his chin. A pair of round-framed glasses pinched the bridge of his nose, and his black wool suit was worn at the edges, the knees worn thin and shiny.

"Ten thousand pardons, Your Majesty," the Lord Chamberlain began, bowing from the waist, "but may I introduce to you the Proctor Napier, scientific attaché to the Imperial Capital from the subjugated land of Britain, conquered in centuries past by your glorious ancestors."

The Emperor inclined his head slightly, indicating that the foreign devil could continue.

"Many thanks for this indulgence, O Emperor," Proctor Napier began. "I come seeking your patronage."

The Emperor twitched the fingers of one hand, a precise motion.

"I was sent to these shores by your servant government in my home island," Napier continued, "to assist in Imperial research. My specialty is logic, and the ordering of information, and over the course of the past years I have become increasingly involved with the questions of computation. The grand designs of Your Majesty's long range plans, whether to explore the moon and far planets, or to chart the course of the stars across the heavens, demand that complex calculations be performed at every step, and each of these calculations require men, materials, and time. It is my hope that each of these three prerequisites might be eliminated to a degree, so as to speed the progress toward your goals."

Tsui, not certain before this moment why he had been called before the Emperor, now harbored a suspicion, and stifled the desire to shout down the foreign devil. At the Chamberlain's side, he listened on, his hands curled into tense fists in his long sleeves.

"With Your Majesty's kind indulgence," Napier said, "I would take a moment to explain the fundaments of my invention." With a timid hand, he gestured toward the oily contraption on the floor behind him. "The basic principle of its operation is a number system of only two values. I call this system *binary.* Though an innovation of Europe, this system has its basis in the ancient wisdom of China, and as such it seems appropriate that Your Divine Majesty is the one to whom it is presented.

"The trigrams of the I Ching are based on the structure of yin and yang, the complementary forces of nature. These trigrams, the building blocks of the I Ching, are composed either of broken or of unbroken lines. Starting from this pair of values, any number of combinations can be generated. Gottfried Leibniz, a German sage, adapted this basic structure some two hundred years ago into a full number system, capable of encoding any value using only two symbols. He chose the Arabic numerals *1* and *0,* but the ideograms for yin and yang can be substituted and the system still functions the same. The decoding is key. Using the Arabic notation, the number one is represented as *1,* the number two as *10,* the number three as *11,* the number four as *100,* and so on."

The Emperor sighed pointedly and glanced to the spacecraft model in his lap, signifying that he was growing weary of the presentation.

"Oh, dear," Napier whispered under his breath, and then hastened to add, "Which brings me to my invention." He turned and stepped to the side of the construct of oily metal and wood on the floor. It was about the height of a man's knee, almost as wide, a roughly cubical shape of copper and iron, plain and unadorned. The top face was surmounted by a brass frame, into which was set a series of wooden blocks, each face of which was carved with a number or symbol. On the cube face presented toward the Emperor was centered an array of articulated brass buttons, three rows of fifteen, the brassy sheen dulled by smudges of oil and grime.

"I call it the Analytical Engine. Powered by a simple motor, the engine comprises a series of switches, each of which can be set either to an 'on' or 'off' state by the manipulation of gears and cogs. By assigning a binary value to each of the two states, we are then able to represent with the engine any numerical value conceivable, so long as there are a sufficient number of switches available. With the inclusion of five operational variables, and the ability to display results immediately"—he indicated the series of blocks crowning the device—"a fully functional Analytical Engine would theoretically be capable of solving quickly any equation put to it. Anyone with a rudimentary ability to read and input values can produce results more quickly and efficiently than a team of trained abacists. This is only a prototype model, of course, capable of working only up to a limited number of digits, but with the proper funding I'm confident we could construct an engine free of this limitation."

Tsui's pulse raged in his ears, though he kept silent and calm in the view of the Emperor.

"If I may?" Napier said, glancing from the Emperor to his invention with an eyebrow raised.

The Emperor twitched, almost imperceptibly, and in response the Chamberlain stepped forward.

"You may exhibit your device," the Chamberlain announced, bowing his head fractionally but never letting his eyes leave Napier's.

Wiping his hands nervously on the thin fabric of his pants, Napier crouched down and gripped the wood-handled crank at the rear of the engine. Leaning in, the strain showing on his pale face, he cranked through a dozen revolutions that produced a grinding clatter that set Tsui's teeth on edge. Finally, when the Chief Computator was sure he could stand the torture no longer, the engine sputtered, coughed, and vibrated to clanking life. Little plumes of acrid smoke billowed up from the corners of the metal cube, and a slow drip of oil from one side puddled in a growing pool on the lacquered floor.

Licking his lips, Napier worked his way around to the front of the device and rested his fingers on the rows of brass buttons. "I'll start with a simple operation," he announced. "Can anyone provide two numbers?"

No one ventured an answer, all too occupied with the clattering machine on the floor, afraid that it might do them some harm.

"You, sir?" Napier said, pointing at Tsui. "Can you provide me with two numbers for my experiment?"

All eyes on him, not least of which the Emperor's, Tsui could only nod, biting back the answer that crouched behind his teeth, hoping to pounce.

"One and two," Tsui answered simply, eyes on the floor.

With a last look around the assembled for any other response, Napier hit four buttons in sequence.

"I've just instructed the engine to compute the sum of the two provided values," he explained, pausing for a brief resigned sigh, "and when I press this final button the calculation will occur immediately and the result will be displayed above."

Demonstrating a flair for the dramatic, Napier reached back his hand and stabbed a finger at the final button with a flourish. The engine smoked and wheezed even more than before, and with a final clatter the rightmost of the blocks crowning the device spun on its brass axis and displayed the symbol for 3 faceup.

"There, you see?" Napier said. "The answer produced, without any human intervention beyond the initial input."

"I have seen horses," the Emperor replied in a quiet voice, "clopping their hoofs on cobblestones, do more complicated sums than this."

"Perhaps, Your Majesty," the Chamberlain said, stepping forward, "a more evaluative demonstration is in order. Chief Computator Tsui?" The Chamberlain motioned to him with a brief wave of his hand, and Tsui inched forward, his fingers laced fiercely together in front of him.

The Chamberlain then snapped his fingers, and a page glided out of the shadows into the center of the hall, a small stool in one hand, an abacus in the other. Setting the stool down a few paces from the foreign devil's instrument, the page presented the abacus to Tsui and, bowing low, glided back into the shadows.

"I would suggest, with Your Majesty's permission," the Chamberlain said, "that a series of calculations be performed, both by the Proctor Napier and his machine, and by our own Chief Computator and his abacus. Which of the two performs more reliably and efficiently will no doubt tell us more than any other demonstration could."

The Emperor twitched his eyebrows, slightly, suggesting a nod.

"Let us begin," said the Chamberlain.

Tsui seated himself on the stool. The abacus on his lap was cool and smooth at his touch, the beads when tested sliding frictionless over the frame of rods. Tilting the frame of the abacus up, he set the beads at their starting position and then left his fingers hovering over the rightmost row, ready to begin.

* * *

The Chamberlain officiated, providing values and operations from a slip of paper he produced from his sleeve. That he'd anticipated this test of man and machine was obvious, though it was inappropriate for any involved to suggest the Chamberlain had orchestrated the events to his ends.

The first calculation was a simple addition, producing the sum of two six-digit numbers. Tsui had his answer while Napier's engine was still sputtering and wheezing, taking less than a third of the time needed for the machine to calculate and display the correct answer on blocks.

The second calculation was multiplication, and here again Tsui finished first. The lapse of time between Tsui calling out his answer and Napier calling out his, though, dwindled in this second round, the engine taking perhaps only twice as long.

The third calculation was division, a four-digit number divided into a six-digit one. Tsui, pulse racing, called out his answer only an instant before Napier. The ruling of the Chamberlain named the Chief Computator the victor, even after Napier protested that he had inadvertently set his engine to calculate to two decimal places, and that as a result his answer was in fact more accurate.

The fourth and final calculation was to find the cube root of a six-digit number. This time, with his previous failure in mind, Napier shouted out after the numbers had been read that the answer should be calculated to two decimal points. The Chamberlain, eyes on the two men, nodded gravely and agreed to this condition. Tsui, who was already fiercely at work on the solution, felt the icy grip of dread. Each additional decimal place in a cube root operation increased the time necessary for the computation exponentially, and even without them he wasn't sure if he would finish first.

Fingers racing over the beads, too tense even to breathe, Tsui labored. The answer was within reach, he knew, with only seconds until he would be named the victor. The abhorrent clattering machine of the foreign interloper would be exposed for a fraud, and the place of the Chief Computator, and of the Imperial House of Computation, would be secure.

"I have it!" Napier shouted, and stepped back from the Analytical Engine to let the assembled see the displayed solution. There was a manic gleam in his eyes, and he looked directly at the Emperor without reservation or shame, as though expecting something like applause.

Tsui was frozen, struck dumb. Reviewing his mental calculations, he realized he'd been nowhere near an answer, and would have required minutes more even to come close. He looked up, saw the symbols displayed on the first blocks of the device, and knew that Napier's answer was the correct one.

"It is decided, then," the Chamberlain announced, striding to Tsui's side. "Of the four tests, the methods of our tradition won out more often than they did not, and only by changing the parameters of the examination after calculations had begun was the Proctor Napier able to prevail. Napier's device is a failure."

"But . . . ," Napier began, on the edge of objection. Seeing the stern expression on the Chamberlain's face, and looking to the palace guards that ringed the room, the foreigner relented. He'd agreed that his machine should be judged by a majority of tests, and he had to abide by the results. To object now would risk a loss of face, at best, and a loss of something much more dire at worst.

Tsui, too numb still to speak, rose shakily to his feet and handed the abacus back to the page, who appeared again from the shadows. Bowing to the Emperor, he backed toward the exit, face burning with self-recrimination..

"The Emperor demands a brief moment," the Emperor announced, sitting forward with something resembling interest. "British, how much time and work would be needed for you to complete the improvements you mentioned earlier? How many of your countrymen are trained in the arts of this device, who could assist you in the process?"

Napier, already in the process of packing up his engine dejectedly, rose to his feet. Rubbing his lower lip with an oil-stained finger, he answered.

"A matter of months to eradicate the current limitations, Your Majesty," he said. "Perhaps a year. But I would need easily as much time to instruct a staff of men, as at present I am the only one who understands all the aspects of the engine's manufacture."

The Emperor, uncharacteristically demonstrative, nodded twice.

"Leave now," the Emperor commanded, and they did.

* * *

In the antechamber, while Napier led a collection of pages and eunuchs in dismantling and boxing up his device, the Chamberlain caught Tsui's elbow.

"A moment, Chief Computator," the Chamberlain said in a low voice, drawing him into an alcove and well out of earshot.

"My thanks, O Lord Chamberlain," Tsui said, his tones hushed, "for allowing me to perform this small service for our master the Emperor."

"We all serve our part," the Chamberlain answered. "Remember, though, that the Emperor's remembrance of this good office will serve only to balance his displeasure that you kept him waiting."

"And for that, you have my apologies," Tsui answered. "But it is strange, I should think, that you would send for me at the House of Computation, in an hour during which it is well known to you that I am elsewhere at my leave. Would not one of my journeymen have been a suitable representative to hear the foreigner's presentation, and to offer any service you might require?"

"Perhaps," the Chamberlain replied, eyes narrowed. "Perhaps it slipped my memory that you would not be found in the House of Computation at this hour, and perhaps it did not occur to me that one of your able journeymen might be as suited for our purposes. But perhaps"—the Chamberlain raised a long finger—"it was best that a member of the House of Computation in your position of leadership was present to see and hear what you have. I have always counted on you, O Chief Computator, to find solutions to problems others thought without resolution. Even, I add, solutions to things others did not even see as problems."

Tsui nodded.

"Yes," he said, "but of the many hundreds who labor under me in the art of calculation, there are others very nearly as adept." He paused, and then added, "Many hundreds."

"Mmm," the Chamberlain hummed. "It is best, then, do you not think, that this device of the British does not meet the Emperor's standards, that so many hundreds of adepts are not removed from their productive positions?"

That the standards proposed had not been the Emperor's, but had instead been proposed by the Lord Chamberlain himself, was a point Tsui did not have to raise. The Emperor, in fact, as evidenced by his uncharacteristic inquiry into the production cycle of Napier's invention, seemed not entirely swayed by the Lord Chamberlain's stagecraft, the question of the utility of the Analytical Engine not nearly so closed as Tsui might have hoped.

"I could not agree more," Tsui answered, thin-lipped and grave. "I thank you for this consideration, and value our exchange."

The Chamberlain nodded and, drawing his robes around him, slid away into the antechamber and beyond, leaving Tsui alone.

The next morning found Tsui in the Ornamental Garden, eyes closed, by the northernmost abacus fishpond.

The noise of shoes scuffing on gravel at his side startled him, and he opened his eyes to see Royal Inspector Bai standing at his side. He'd made no other sound in his approach.

"Good morning, Chief Computator," Bai said, a statement more than a question.

"Yes, Inspector," answered Tsui, looking down into the waters of the pond. They were silty and gray, the carnivorous fish almost hidden below the surface. "I would say that it is."

"Surprising, one might argue," Bai went on, "after the excitement of the evening." The Inspector pulled a waxed-paper-wrapped lump of meat and bread from within his sleeve and, unwrapping it, began to drop hunks of dried pork into the waters.

"Excitement?" Tsui asked innocently.

"Hmm," the Inspector hummed, peering down into the water, which was quiet and still but for the ripples spreading out from the points where the meat had passed. "The fish seem not very hungry today," he said softly, distracted, before looking up and meeting Tsui's gaze. "Yes," he answered, "excitement. It seems that a visitor to the Forbidden City, a foreign inventor, went missing somewhere between the great hall and the main gate after enjoying an audience with the Emperor. The invention that he'd brought with him was found scattered in pieces in the Grand Courtyard, the box that held it appearing to have been dropped from a high-story balcony, though whether by accident or design we've been unable to determine. The Emperor has demanded the full attentions of my bureau be trained on this matter, as it seems that he had some service with which to charge this visitor.

That the visitor is not in evidence, and this service might go unfilled, has done little to improve the temper of our master, equal-of-heaven and may-he-reign-ten-thousand-years."

Tsui nodded, displaying an appropriate mixture of curiosity and concern.

"As for the man himself," Bai said, shrugging, "as I've said, he seems just to have vanished." The Inspector paused again and, in a practiced casual tone, added, "I believe you were present at the foreign inventor's audience yesterday, yes? You didn't happen to see him at any point following his departure from the hall, did you?"

Tsui shook his head, and in all sincerity answered, "No."

The Chief Computator had no fear. He'd done nothing wrong, after all, his involvement in the business beginning with a few choice words to his more perceptive journeymen and foremen on his hurried return to the Imperial House of Computation and ending in the early morning hours when a slip of paper was delivered to him by one of his young apprentices. On the slip of paper, unsigned or marked by any man's chop, was a single ideogram, indicating *Completion* but suggesting *Satisfaction.*

Tsui's business, since childhood, had been identifying problems and presenting solutions. To what uses those solutions might be put by other hands was simply not his concern.

"Hmm," the Inspector hummed again, and looking at the still waters of the pond, he shook his head. "The abacus fish just don't seem interested today in my leavings. Perhaps they've already been fed, yes?"

"Perhaps," Tsui agreed.

The Inspector, with a resigned sigh, dropped the remainder of the meat into the northernmost pond and then tossed the remaining bread into the southernmost, where the languid fish began their slow ballet to feed themselves.

"Well, the Emperor's service demands my attention," Inspector Bai said, brushing off his hands, "so I'll be on my way. I'll see you tomorrow, I trust?"

Tsui nodded. "Yes," he answered, "I don't expect that I'll be going anywhere."

The Inspector gave a nod, which Tsui answered with a slight bow, and then left the Chief Computator alone in the garden.

Tsui looked down into the pond and saw that the silt was beginning to settle on the murky bottoms, revealing the abacus fish arranged in serried ranks, marking out the answer to some indefinable question. The Chief Computator closed his eyes, and in the silence imagined countless men working countless abaci, tirelessly. His thoughts on infinity, Tsui smiled.

Over the past twenty years, the prolific and wildly creative **Paul Di Filippo** has published well over one hundred stories, distributed across nearly every SF magazine. Collections of his work include *The Steampunk Trilogy, Ribofunk, Fractal Paisleys, Lost Pages,* and *Little Doors.* He lives in Providence, Rhode Island, with his mate, Deborah Newton.

CLOUDS AND COLD FIRES

Paul Di Filippo

Out of a clear sky on a fine summer morning, a buckshot rattle of hailstones across the living pangolin plates of Pertinax's rooftop announced the arrival of some mail.

Inside his cozy, low-ceilinged hutch, with its corner devoted to an easel and canvases and art supplies, its shelves full of burl sculptures, its workbench that hosted bubbling retorts and alembics and a universal proseity device, Pertinax paused in the feeding of his parrot tulips. Setting down the wooden tray of raw meat chunks, he turned away from the colorfully enameled soil-filled pots arrayed on his bright windowsill. The parrot tulips squawked at this interruption of their lunch, bobbing their feathery heads angrily on their long succulent neck stalks. Pertinax chided them lovingly, stroking their crests while avoiding their sharp beaks. Then, hoisting the hem of his long striped robe to expose his broad naked paw-feet, he hurried outdoors.

Fallen to the earth after bouncing from the imbricated roof, the hailstones were already nearly melted away to invisibility beneath the temperate sunlight, damp spots on the undulant greensward upon which Pertinax's small but comfortable dwelling sat. Pertinax wetted a finger, raised it to gauge the wind's direction, and then directed his vision upward and to the north, anticipating the direction from which his mail would arrive. Sure enough, within a minute, a lofty cloud had begun to form, a flocculent painterly smudge on the monochrome canvas of the turquoise sky.

The cloud assumed coherence and substance, drawing into itself its necessary share of virgula and sublimula omnipresent within the upper atmosphere. After another minute or two, the cloud possessed a highly regular oval outline and had descended to within five meters of the ground. Large as one of the windows in Pertinax's hutch, the cloud halted its progress at this level, and its surface began to acquire a sheen. The sheen took on the qualities of an ancient piece of translucent plastic, such as the Overclockers might cherish. Then Pertinax's animated mail appeared across the cloud's surface, as the invisible components of the cloud churned in coordinated fashion.

Sylvanus's snouty whiskered face smiled, but the smile was grim, as was his voice resonating from the cloud's fine-grain speakers:

"Pertinax, my friend, I regret this interruption of your studies and recreations, but I have some dramatic news requiring our attention. It appears that the Overclockers at their small settlement known as 'Chicago' are about to launch an assault on the tropospherical mind. Given their primitive methods, I doubt they can inflict permanent damage. But their mean-spirited sabotage might very well cause local disruptions before the mind repairs itself. I know you have several projects running currently, and I would hate to see you lose any data during a period of limited chaos. I would certainly regret any setbacks to my ongoing modeling of accelerated hopper embryogenesis. Therefore, I propose that a group of those wardens most concerned form a delegation to visit the Overclockers and attempt to dissuade them from such malicious tampering. Mumbaugh has declined to participate—he's busy dealing with an infestation of hemlock mites attacking the forests of his region—but I have firm committments from Cimabue, Tanselle, and Chellapilla. I realize that it is irksome to leave behind the comforts of your home to make such a trip. But I am hoping that I may count on your participation, as well. Please reply quickly, as time is of the essence."

Its mail delivered, the cloud wisped away into its mesoscopic constituent parts. A light misty drizzle refreshed Pertinax's face. But otherwise he was left with only the uneasy feelings occasioned by the message.

Of course he would help Sylvanus. Interference with the tropospherical mind could not be tolerated. The nerve of those Overclockers!

Not for the first time, nor probably the last, Pertinax ruefully contemplated the dubious charity of the long-departed Upflowered.

When 99.9 percent of humanity had abandoned the Earth for greener intergalactic pastures during the Upflowering, the leavetakers had performed several final tasks. They had rearcadized the whole globe, wiping away nearly every vestige of mankind's crude twenty-second-century protocivilization, and restocked the seas and plains with many beasts. They had established Pertinax and his fellows—a small corps of ensouled, spliced, and redacted domestic animals—as caretakers of the restored Earth. They had charitably set up a few agrarian reservations for the small number of dissidents and malfunctioning humans who chose to remain behind, stubbornly unaltered in their basic capabilities from their archaic genetic baseline. And they had uploaded every vestige of existing machine intelligence and their knowledge bases to a new platform: an airborne network of minuscule, self-replenishing components, integrated with the planet's meteorological systems.

During the intervening centuries, the remaining archaic humans—dubbed the Overclockers for their uncanny devotion to both speed and the false quantization of holistic imponderables—had gradually dragged themselves back up to a certain level of technological achievement. Now, it seemed, they were on the point of making a nuisance of themselves. This could not be tolerated.

Hurrying back into his compact domicile, Pertinax readied his reply to Sylvanus. From a small door inset in one wall, which opened onto a coop fixed to the outside wall, Pertinax retrieved a mail pigeon. He placed the docile murmuring bird on a tabletop and fed it some special seed scooped from one compartment of a feed bin. While he waited for the virgula and sublimula within the seed to take effect, Pertinax supplied his own lunch: a plate of carrots and celery, the latter smeared with delicious bean paste. By the time Pertinax had finished his repast, cleaning his fur with the side of one paw-hand all the way from muzzle to tufted ear tips, the pigeon was locked into recording mode, staring ahead fixedly, as if hypnotized by a predator.

Pertinax positioned himself within the bird's field of vision. "Sylvanus, my peer, I enlist wholeheartedly in your mission! Although my use of the tropospheric mind is negligible compared with your own employment of the system, I do have all my statistics and observations from a century of avian migrations stored there. Should the data and its backups be corrupted, the loss of such a record would be disastrous! I propose to set out immediately by hopper for Chicago. Should you likewise leave upon receipt of this message, I believe our paths will intersect somewhere around these coordinates." Pertinax recited latitude and longitude figures. "Simply ping my hopper when you get close enough, and we'll meet to continue the rest of our journey together. Travel safely."

Pertinax recited the verbal tag that brought the pigeon out of its trance. The bird resumed its lively attitude, plainly ready to perform its share of the mail delivery. Pertinax cradled the bird against his oddly muscled chest and stepped outside. He lofted the pigeon upward, and it began to stroke the sky bravely.

Once within the lowest layers of the tropospheric mind, the bird would have its brain states recorded by an ethereal cap of spontaneously congregating virgula and sublimula, and the bird would be free to return to its coop.

Pertinax's message would thus enter the meterological medium and be propagated across the intervening leagues to Sylvanus. Like a wave in the ocean, the information was not dependent upon any unique set of entities to constitute its identity, and so could travel faster than simple forward motion of particles might suggest. To span the globe from Pertinax to the antipodes took approximately twelve hours, and Sylvanus lived much closer. Not as fast as the ancient quantum-entanglement methods extant in the days before the Upflowering. But then again, the pace of life among the stewards was much less frenzied than it had been among the ancestors of the Overclockers.

Having seen his mail on its way, Pertinax commenced the rest of his preparations for his trip. He finished feeding his parrot tulips, giving them a little extra to see them through his time away from home. If delayed overlong, Pertinax knew they would estivate safely till his return. Then from a cupboard he took a set of large saddlebags. Into these pouches he placed victuals for himself and several packets of multipurpose pigeon seed, as well as a few treats and vitamin pills to supplement the forage that his hopper would subsist on during the journey. He looked fondly at his neat, comfortable bed, whose familiar refuge he would miss. No taking that, of course! But the hopper would provide a decent alternative. Pertinax added a few other miscellaneous items to his pack, then deemed his provisions complete.

Stepping outside, Pertinax took one fond look back inside before shutting and latching his door. He went around shuttering all the windows as a precaution against the storms that sometimes accompanied the more demanding calculations of the tropospheric mind. From the pigeon coop he withdrew three birds and placed them in a loosely woven wicker carrier. Then he took a few dozen strides to the hopper corral, formed of high walls of living ironthorn bush.

Pertinax's hopper was named Flossy, a fine mare. The redacted Kodiak kangemu stood three meters tall at her shoulders. Her pelt was a curious blend of chestnut fur and gray feathers; her fast-twitch-muscled legs were banded with bright yellow scales along the lower third above her enormous feet. A thick strong tail jutted backwards for almost half Flossy's length.

Pertinax tossed Flossy a treat, which she snapped from the air with her long jaws. In the stable attached to the corral, Pertinax secured a saddle. This seat resembled a papoose or backpack, with two shoulder straps. Outside again, Pertinax opened the corral gate—formed of conventional timbers—and beckoned to Flossy, who obediently came out and hunkered down. Holding the saddle up above his head, Pertinax aided Flossy in shrugging into the seat. He cinched the straps, then hung his saddlebags from one lower side of the seat and the wicker basket containing the pigeons from the other. Deftly Pertinax scrambled up, employing handholds of Flossy's fur, and ensconced himself comfortably, the seat leaving his arms free but cradling his back and neck. His head was now positioned above Flossy's, giving him a clear view of his path. He gripped Flossy's big upright ears firmly yet not harshly and urged his mount around to face northeast.

"Gee up, Flossy," said Pertinax, and they were off.

Flossy's gait was the queerest mixture of hopping, vaulting, running, and lumbering, a mode of locomotion unknown to baseline creation. But Pertinax found it soothing, and his steed certainly ate up the kilometers.

For the first few hours, Pertinax enjoyed surveying his immediate territory, quite familiar and beloved, noting subtle changes in the fauna and flora of the prairie that distance brought. In early afternoon he stopped for a meal, allowing Flossy to forage. Taking out a pigeon and prepping the bird, Pertinax recited his morning's scientific observations to be uploaded to the tropospheric mind. Its data delivered, the bird homed back to Pertinax rather than the cottage. In less than an hour, the warden was under way again.

Pertinax fell asleep in the saddle and awoke at dusk. He halted Flossy and dismounted to make camp. With the saddle off, Flossy cropped wearily nearby. The first thing Pertinax attended to was the establishment of a security zone. A pheromonal broadcaster would disseminate the warden's exaggerated chemical signature for kilometers in every direction, a note that all of wild creation was primed by the Upflowered to respond to. Avoidance of the distinctive trace had been built into their ancestors' genes. (The bodily signature had to be masked for up close work with animals.) Pertinax had no desire to be trampled in the night by a herd of bison, or attacked by any of the region's many predators. Sentient enemies were nonexistent, with the nearest Overclockers confined by their limited capacities nearly one thousand kilometers away in Chicago.

After setting up the small scent-broadcast unit, Pertinax contemplated summoning forth some entertainment. But in the end he decided he was just too tired to enjoy any of the many offerings of the tropospheric mind, and that he would rather simply go to sleep.

The upright Flossy, balanced tripodally on her long tail, was already herself half a-drowse, and she made only the softest of burblings when Pertinax clambered into her capacious marsupial pouch. Dry and lined with a soft down, the pouch smelled like the nest of some woodland creature, and Pertinax fell asleep feeling safe and cherished.

The morning dawned like the first day of the world, crisp and inviting. Emerging from his nocturnal pouch, Pertinax noted that night had brought a heavy dew that would have soaked him had he been dossing rough. But instead he had enjoyed a fine, dry, restful sleep.

Moving off a ways from the grumbling Flossy and casting about with a practiced eye, Pertinax managed to spot some untended prairie chicken nests amidst the grassy swales. He robbed them of an egg apiece without compunction (the population of the birds was robust), and soon a fragrant omelette, seasoned with herbs from home, sizzled over a small propane burner. (Pertinax obtained the flammable gas from his universal proseity device, just as he supplied many of his needs.)

After enjoying his meal, Pertinax dispatched a pigeon upward to obtain from the tropospheric mind his positional reading, derived from various inputs such as constellational and magnetic. The coordinates, cloud-blazoned temporarily on the sky in digits meters long, informed Pertinax that Flossy had carried him nearly 150 kilometers during their previous half-day of travel. At this rate, he'd join up with Sylvanus on the morrow, and with the others a day later. Then the five stewards would reach Chicago around noon of the fourth day.

Past that point, all certainty vanished. How the Overclockers would react to the arrival of the wardens, how the wardens would dissuade the humans from tampering with the planetary mind, what they would do if they met resistance—all this remained obscure.

Remounting Flossy, Pertinax easily put the uncertainty from his mind. Neither he nor his kind were prone to angst. So, once on his way, he reveled instead in the glorious day and the unfolding spectacle of a nature reigning supreme over an untarnished globe.

Herds of bisons thundered past at a safe distance during various intervals along Pertinax's journey. Around noon a nearly interminable flock of passenger pigeons darkened the skies. A colony of prairie dogs stretching across hectares mounted a noisy and stern defense of their town.

That night replicated the simple pleasures of the previous one. Before bedding down, Pertinax enjoyed a fine display of icy micrometeorites flashing into the atmosphere. The Upflowered had arranged a regular replenishment of Earth's water budget via this cosmic source before they left.

Around noon on the second full day of travel, with the landscape subtly changing as they departed one bioregion for another, Pertinax felt a sudden quivering alertness thrill through Flossy. She had plainly pinged the must of Sylvanus's steed (a stallion named Bix) on the wind, and needed no help from her rider to zero in on her fellow Kodiak kangemu. Minutes later, Pertinax himself espied Sylvanus and his mount, a tiny conjoined dot in the distance.

Before long, the two wardens were afoot and clasping each other warmly, while their hoppers boxed affectionately at each other.

"Pertinax, you're looking glossy as a foal! How I wish I were your age again!"

"Nonsense, Sylvanus, you look splendid yourself. After all, you're far from old. A hundred and twenty-nine last year, wasn't it?"

"Yes, yes, but the weary bones still creak more than they did when I was a young buck like you, a mere sixty-eight. Some days I just want to drop my duties and retire. But I need to groom a successor first. If only you and Chellapilla—"

Pertinax interrupted his elder friend. "Perhaps Chellapilla and I have been selfish. I confess to feeling guilty about this matter from time to time. But the demands on our energies seemed always to preclude parentage. I'll discuss it with her tomorrow. And don't forget, there's always Cimabue and Tanselle."

Sylvanus clapped a hearty paw-hand on Pertinax's shoulder. "They're fine stewards, my boy, but I had always dreamed of your child stepping into my shoes."

Pertinax lowered his eyes. "I'm honored, Sylvanus. Let me speak of this with Chell."

"That's all I ask. Now I suppose we should be on our way again."

It took some sharp admonishments and a few coercive treats to convince Bix and Flossy to abandon their play for the moment and resume travel, but eventually the two wardens again raced northeast toward their unannounced appointment with the Overclockers.

That night before turning in, Sylvanus suggested some entertainment.

"I have not viewed any historical videos for some time now. Would you care to see one?"

"Certainly. Do you have a suggestion?"

"What about *The Godfather*?"

"Part One?"

"Yes."

"An excellent choice. Perhaps it will help to refresh our understanding of Overclocker psychology. I'll send up a pigeon."

The sleepy bird responded sharply to the directorial seed and verbal instructions, then zoomed upward. While the wardens waited for the tropospheric mind to respond, they arranged their packs and saddles in a comfortable couch that allowed them to lie back and observe the nighted skies.

In minutes a small audio cloud had formed low down near them, to provide the soundtrack. Then the high skies lit with colored cold fires.

The new intelligent meterology allowed for auroral displays at any latitude of the globe, as cosmic rays were channeled by virgula and sublimula, then bent and manipulated to excite atoms and ions. Shaped and permuted on a pixel level by the distributed airborne mind, the auroral canvas possessed the resolution of a twentieth-century drive-in screen, and employed a sophisticated palette.

Clear and bold as life, the antique movie began to unroll across the black empyrean. Snacking on dried salted crickets, the two stewards watched in rapt fascination until the conclusion of the film.

"Most enlightening," said Sylvanus. "We must be alert for such incomprehensible motives as well as deceptions and machinations among the Overclockers."

"Indeed, we would be foolish to anticipate any rationality at all from such a species. Their ancestors' choice to secede from the Upflowering tells us all we need to know about their unchanged mentality."

Midafternoon of the next day found Sylvanus and Pertinax hard-pressed to restrain their rambunctious hoppers from charging toward three other approaching Kodiak kangemu. At the end of the mad gallop, five stewards were clustered in a congregation of hearty backslapping and embraces, while the frolicking hoppers cavorted nearby.

After the general exchange of greetings and reassurances, Cimabue and Tanselle took Sylvanus one side to consult with him, leaving Pertinax and Chellapilla some privacy.

Chellapilla smiled broadly, revealing a palisade of blunt healthy brown teeth. Her large hazel eyes sparkled with affection, and her leathery nostrils flared wetly. The past year since their last encounter had seen her acquire a deep ragged notch in one ear. Pertinax reached up to touch the healed wound. Chellapilla only laughed, before grabbing his paw-hand and kissing it.

"Are you troubled by that little nick, Perty? Just a brush with a wounded wolverine when I was checking a trapline for specimens last winter. Well worth the information gained."

Pertinax found it hard to reconcile himself to Chellapilla's sangfroid. "I worry about you, Chell. It's a hard life we have sometimes, as isolated guardians of the biosphere. Don't you wish, just once in a while, that we could live together? . . ."

"Ah, of course I do! But where would that end? Two stewards together would become four, then a village, then a town, then a city of wardens. With our long lifespans, we'd soon overpopulate the world with our kind. And then Earth would be right back where it was in the twenty-second century."

"Surely not! Our species would not fall prey to the traps mankind stumbled into before the Upflowering."

Chellapilla smiled. "Oh, no, we'd be clever enough to invent new ones. No, it's best this way. We have our pastoral work to occupy our intelligence, with the tropospheric mind to keep us in daily contact and face-to-face visits at regular intervals. It's a good system."

"You're right, I suppose. But still, when I see you in the flesh, Chell, I long for you so."

"Then let's make the most of this assignment. We'll have sweet memories to savor when we part."

Pertinax nuzzled Chellapilla's long furred neck, and she shivered and clasped him close. Then he whispered his thoughts regarding Sylvanus's desired retirement and the needful successor child into her ear.

Chellapilla chuckled. "Are you sure you didn't put Sylvanus up to this? You know the one exemption from cohabitation is the period of parenting. This is all a scheme to get me to clean your hutch and cook your meals on a regular basis for a few years, isn't it?"

"Yes, I admit it. There's never been a universal proseity device made that was as nice to hold as you."

"Well, let me think about it for the rest of this trip, before I go off my pills. It's true that you and I are not getting any younger, and I am inclined toward becoming a mother, especially if our child will help ease Sylvanus's old age. But I want to make sure I'm not overlooking any complications."

"My ever-sensible Chell! I could have dictated your reply without ever leaving my hutch."

Chellapilla snorted. "One of us has to be the sobersided one."

The two lovers rejoined their fellow stewards. Tanselle immediately took Chellapilla to one side, in an obvious attempt to pump her friend for any gossip. The feminine whispers and giggles and sidewise glances embarrassed Pertinax, and he made a show of engaging

Cimabue in a complex discussion of the latter's researches. But Pertinax could lend only half his mind to Cimabue's talk of fisheries and turtle breeding, ocean currents and coral reefs. The other half was still contemplating his exciting future with Chellapilla.

Eventually Sylvanus roused them from their chatter with a suggestion that they resume their journey. Bix, Flossy, Amber, Peavine, and Peppergrass bore their riders north, deeper into the already encroaching forests of the Great Lakes region.

When they established camp that evening in a clearing beneath a broad canopy of lofty treetops, Sylvanus made a point of setting up a little hearth somewhat apart from his younger comrades. Plainly, he did not want to put a damper on any romantic moments among the youngsters.

The five shared supper together, however. Sylvanus kept wrinkling his grizzled snout throughout the meal, until finally he declaimed, "There's a storm brewing. The tropospheric mind must be performing some large randomizations or recalibrations. I suspect entire registers will be dumped."

Baseline weather had been tempered by the creation of an intelligent atmosphere. Climates across the planet were more equitable and homogenous, with fewer extreme instances of violent weather. But occasionally both the moderately large and even the titanic disturbances of yore would recur, as the separate entities that constituted the community of the skies deliberately encouraged random Darwinian forces to cull and mutate their members.

"I packed some tarps and ropes," said Cimabue, "for just such an occasion. If we cut some poles, we can erect a shelter quickly."

Working efficiently, the wardens built, first, a three-walled roofed enclosure for their hardy hoppers, stoutly braced between several trees, its open side to the leeward of the prevailing winds. Then they fashioned a small but sufficient tent for themselves and their packs, heavily staked to the earth. A few blankets strewn about the interior created a comfy nest, illuminated by several cold luminescent sticks. Confined body heat would counter any chill.

Just as they finished, a loud crack of thunder ushered in the storm. Safe and sound in their tent, the wardens listened to the rain hammering the intervening leaves above before filtering down to drip less heavily on their roof.

Sylvanus immediately bade his friends good night, then curled up in his robe in one corner, his back to them. Soon his snores—feigned or real—echoed off the sloping walls.

Swiftly disrobed, Cimabue and Tanselle began kissing and petting each other, and Pertinax and Chellapilla soon followed suit. By the time the foursome had begun exchanging cuds, their unashamed mating, fueled by long separation, was stoked to proceed well into the night.

The reintegrational storm blew itself out shortly after midnight, with what results among the mentalities of the air the wardens would discover only over the course of many communications. Perhaps useful new insights into the cosmos and Earth's place therein had been born this night.

In the morning the shepherds broke down their camp, breakfasted, and embarked on the final leg of their journey to Chicago. Pertinax rode his hopper in high spirits, pacing Chellapilla's Peavine.

Not too long after their midday meal (Tanselle had bulked out their simple repast with some particularly tasty mushrooms she had carried from home), they came within sight of the expansive lake, almost oceanic in its extent, that provided the human settlement with water for both drinking and washing, as well as various dietary staples. Reckoning themselves a few dozen kilometers south of the humans, the five headed north, encountering large peaceful herds of elk and antelopes along the way.

They smelled Chicago before they saw it.

"They're not burning petroleum, are they?" asked Cimabue.

"No," said Sylvanus. "They have no access to any of the few remaining played-out deposits of that substance. It's all animal and vegetable oils, with a little coal from near-surface veins."

"It sure does stink," said Tanselle, wrinkling her nose.

"They still refuse our offer of limited universal proseity devices?" Pertinax inquired.

Sylvanus shook his head ruefully. "Indeed. They are stubborn, suspicious, and prideful, and disdain the devices of the Upflowered as something near-demonic. They claim that such cornucopia would render their species idle and degenerate, and destroy their character. When the Upflowered stripped them of their twenty-second-century technology, the left-behind humans conceived a hatred of their ascended brethren. Now they are determined to reclimb the same ladder of technological development they once negotiated, but completely on their own."

Cimabue snorted. "It's just as well they don't accept our gifts. The UPDs would allow them to spread their baneful way of life even farther than they already have. We can only be grateful their reproductive rates have been redacted downward."

"Come now," said Chellapilla, "surely the humans deserve as much respect and right to self-determination as any other species. Would you cage up all the blue jays in the world simply because they're noisy?"

"You don't have any humans in your bioregion, Chell. See what you think after you've met them."

The pathless land soon featured the start of a crude gravel-bedded road. The terminus the travelers encountered was a dump site. The oil-stained ground, mounded with detritus both organic and manufactured, repelled Pertinax's sensibilities. He wondered how the humans could live with such squalor, even on the fringes of their settlement.

Moving swiftly down the pebbled roadway, the wardens soon heard a clanking, chugging, ratcheting riot of sound from some ways around the next bend of the tree-shaded road. They halted and awaited the arrival of whatever vehicle was producing the clamor.

The vehicle soon rounded the curve of road, revealing itself to be a heterogenous assemblage of wood and metal. The main portion of the carrier was a large wooden buckboard with two rows of seats forward of a flatbed. In the rear, a large boiler formed of odd-shaped scavenged metal plates threatened to burst its seams with every puff of smoke. Transmission of power to the wheels was accomplished by whirling leather belts running from boiler to wood-spoked iron-rimmed wheels.

Four men sat on the rig, two abreast. Dressed in homespun and leathers, they sported big holstered sidearms. The guns were formed of ceramic barrels and chambers, and carved grips. Small gasketed pump handles protruded from the rear of each gun. Pertinax knew the weapons operated on compressed air and fired only nonexplosive projectiles. Still, sometimes the darts could be poison-tipped. A rack of rifles of similar construction lay within easy reach. The driver, busy with his tiller-style steering mechanism and several levers, was plainly a simple laborer. The other three occupants seemed dignitaries of some sort. Or so at least Pertinax deduced, judging from various colorful ribbons pinned to their chests and sashes draped over their shoulders.

Surprised by the solid rank of mounted wardens looming high over the car like a living wall across the dump road, the Overclockers reacted with varying degrees of confusion. But soon the driver managed to bring his steamcart to a halt, and the three officials had regained a measure of diplomatic aplomb. The passenger in the front seat climbed down and aproached Pertinax and friends. Leery of the stranger, the Kodiak kangemus unsheathed their long thick claws a few inches. The awesome display brought the man to a halt a few meters away.

He spoke, looking up and shielding his eyes against the sun. "Hail, wardens! My name is Brost, and these comrades of mine are Kemp and Sitgrave, my assistants. As the mayor of Chicago, I welcome you to our fair city."

Pertinax studied Brost from above, seeing a poorly shaven, sallow baseline *Homo sapiens* with a shifty air about his hunched shoulders. Some kind of harsh perfume failed to mask completely a fug of fear and anxiety crossing the distance between Brost and Pertinax's sharp nostrils.

Sylvanus, as eldest, spoke for the wardens. "We accept your welcome, Mayor Brost. But I must warn you that we are not here for any simple cordial visit. We have good reason to believe that certain factions among your people are planning to tamper with the tropospheric mind. We have come to investigate, and to remove any such threats we may discern."

The mayor smiled uneasily, while his companions fought not to exchange nervous sidelong glances among themselves.

"Tamper with those lofty, serene intelligences, who concern themselves not at all with our poor little lives? What reason could we have for such a heinous assault? No, the charge is ridiculous, even insulting. I can categorically refute it here and now. Your mission has been for naught. You might as well save yourself any further wearisome journeying by camping here for the night before heading home. We will bring you all sorts of fine provisions—"

"That cannot be. We must make our own investigations. Will you allow us access to your village?"

Mayor Brost huddled with his assistants, then faced the wardens again. "As I said, the *city* of Chicago welcomes you, and its doors are open."

Pertinax repressed a grin at the mayor's emphasis on *city,* but he knew the other wardens had caught this token of outraged human dignity, as well.

With much back-and-forward-and-back manuevering, the driver finally succeeeded in turning around the steamcart. Matching the gait of their hoppers to the slower passage of the cart, the wardens followed the delegation back to Chicago.

Beginning with outlying cabins where half-naked childen played in the summer dust of their yards along with mongrels and livestock, and continuing all the way to the "city" center, where a few larger buildings hosted such establishments as blacksmiths, saloons, public kitchens, and a lone bath house, the small collection of residences and businesses that was Chicago—scattered along the lake's margins according to no discernible scheme—gradually assembled itself around the newcomers. Mayor Brost, evidently proud of his domain, pointed out sights of interest as they traversed the "urban" streets, down the middles of which flowed raw sewage in ditches.

"You see how organized our manufactories are," said Brost, indicating some long, low, windowless sheds flanked by piles of waste byproducts: wood shavings, coal clinkers, metal shavings. "And here's the entrance to our mines." Brost pointed to a shack that sheltered a pitlike opening descending into the earth at a slant.

"Oh," said Cimabue, "you're smelting and refining raw metals these days?"

Mayor Brost exhibited a sour chagrin. "Not yet. There's really no need. We feel it's most in harmony with, uh, our beloved mother earth to recycle the buried remnants of our ancestors' civilization. There's plenty of good metal and plastic down deep where the Upflowered sequestered the rubble they left after their redesign of the globe. Plenty for everyone."

"And what exactly is your population these days, Mayor?" inquired Tanselle.

"Nearly five thousand."

Tanselle shook her head reflectively, as if to say, thought Pertinax, *Would that it were even fewer.*

After some additional civic boosterism, the party—considerably enlarged by various gawking hangers-on—arrived at a large, grassy town square, where goats and sheep grazed freely. Ranked across the lawn, tethered securely, were several small lighter-than-air balloons with attached gondolas of moderate size. The shiny laquered patchwork fabric of the balloons lent them a circus air belied by the solemn unease which the mayor and his cohorts eyed the balloons.

Immediately, Pertinax's ears pricked forward at this unexpected sight and the humans' nervous regard for the objects. "What are these for?" he asked.

Mayor Brost replied almost too swiftly. "Oh, these little toys have half a dozen uses. We send up lightweight volunteers to spy out nearby bison herds so that our hunting parties will save some time and trouble. We make surveys from the air for our road-building. And of course, the children enjoy a ride now and then. The balloons won't carry much more weight than a child."

"I'd like to examine them."

"Certainly."

Pertinax clambered down off Flossy. Standing among the humans, the top of his head just cleared their belt buckles. He was soon joined by his fellow wardens, who moved through the crowd like a band of determined furry dwarves.

The balloons featured no burners to inflate their straining shapes. Pertinax inquired as to their source of gas.

Highlighting the mechanisms, Mayor Brost recited proudly. "Each balloon hosts a colony of methanogenic bacteria and a food supply. Increasing the flow of nutrients makes more gas. Closing the petcocks shuts them down."

Pertinax stepped back warily from his close-up inspection of the balloons. "They're highly explosive, then."

"I suppose. But we maintain adequate safety measures around them."

The wardens regrouped off to one side and consulted quietly among themselves.

"Any explosion of this magnitude in the tropospheric mind would do no more damage than a conventional rain squall," said Cimabue.

"Agreed," said Chellapilla. "But what if the explosion was meant to disperse some kind of contaminant carried as cargo?"

"Such as?" asked Tanselle.

"No suitably dangerous substance occurs to me at the moment," Sylvanus said, stroking his chin whiskers.

"Nonetheless," cautioned Pertinax, "I have a feeling that here lies the danger facing the tropospheric mind. Let us continue our investigations for the missing part of the puzzle."

Pertinax returned to address the mayor. "Our mounts need to forage, while we continue our inspection of your town. We propose to leave them here on the green. They will not bother people or livestock, but you should advise your citizens not to molest them. The kangemu are trained to deal harshly with threats to themselves or their masters."

"There will, of course, be no such problems," said the mayor.

Sylvanus advised splitting their forces into two teams for swifter coverage of the human settlement, while he himself, in deference to his age and tiredness, remained behind with their mounts to coordinate the searching. Naturally, Pertinax chose to team up with Chellapilla.

The subsequent hours found Pertinax and his lover roaming unhindered through every part of the human village. Most of the citizens appeared friendly, although some exhibited irritation or a muted hostility at the queries of the wardens. Pertinax and Chellapilla paused only a few minutes to bolt down some cold food around midafternoon before continuing their so-far fruitless search.

Eventually they found themselves down by some primitive docks, watching the small fishing fleet of Chicago tie up for the evening. The fishermen, shouldering their day's bounty in woven baskets, moved warily past the weary wardens.

"Well, I'm stumped," confessed Pertinax. "If they're hiding something, they've concealed it well."

Chellapilla said, "Maybe we're going about this wrong. Let's ask what could harm the virgula and sublimula, instead of just expecting to recognize the agent when we see it."

"Well, really only other virgula and sublimula, which of course the humans have no way of fashioning."

"Ah, but what of rogue lobes?"

The natural precipitation cycles brought infinite numbers of virgula and sublimula down from their habitats in the clouds to ground level. When separated from the tropospheric mind in this way, the components of the mind were programmed for apoptosis. But occasionally a colony of virgula and sublimula would fail to self-destruct, instead clumping together into a rogue lobe. Isolated from the parent mind, the lobes frequently went insane before eventually succumbing to environmental stresses. Sometimes, though, a lobe could live a surprisingly long time if it found the right conditions.

"Do you think local factors in the lake here might encourage lobe formations?"

"There's one way to find out," answered Chellapilla.

It took only another half hour of prowling the lakeshore, scrambling over slippery rocks and across pebbled strands, to discover a small lobe.

Thick intelligent slime latticed with various organic elements—pondweeds, zebra mussels, a disintegrating bird carcass—lay draped across a boulder, a mucosal sac with the processing power of a nonautonomous twenty-second-century AI. The slime was liquescently displaying its mad internal thoughts just as a mail cloud did: fractured images of the natural world, blazes of equations, shards of old human culture ante-Upflowering, elaborate mathematical constructions. A steady whisper of jagged sounds, a schizophrenic monologue, accompanied the display.

Pertinax stared, horrified. "Uploading this fragment of chaos to the tropospheric mind would engender destabilizing waves of disinformation across the skies. The humans don't even need to explode their balloons. Simply letting the mind automatically read the slime would be enough."

"We can't allow this to happen."

"Let's hurry back to the others."

"You damned toothy ratdogs aren't going anywhere."

A squad of humans had come stealthily upon Pertinax and Chellapilla while their attentions were engaged by the lobe. With rifles leveled at their heads, the wardens had no recourse but to raise their hands in surrender.

Two men came to bind the wardens. The one dealing with Chellapilla twisted her arms cruelly behind her, causing her to squeal. Maddened by the sound, Pertinax broke free and hurled himself at one of the gunbearers. But a rifle stock connected with his skull, and he knew only blackness.

When Pertinax awoke, night had fallen. He found himself with limbs bound, lying in a cage improvised from thick branches rammed deep into the soil and lashed together. He struggled to rise, and thus attracted the attentions of his fellow captives.

Similarly bound, Chellapilla squirmed across the grass to her mate. "Oh, Perty, I'm so glad you're awake! We were afraid you had a concussion."

"No, I'm fine. And you?"

"Just sore. Once you were knocked out, they didn't really hurt me further."

Sylvanus's sad voice reached Pertinax, as well. "Welcome back, my lad. We're in a fine mess now, and it's all my fault for underestimating the harmful intentions of these savages."

Firelight flared up some meters away, accompanied by the roar of a human crowd. "Where are we?"

"We're on the town green," said Chellapilla. "The humans are celebrating their victory over us. They slaughtered our kangemu and are roasting them for a feast."

"Barbarians!"

Tanselle spoke. "Cimabue and I are here, as well, Pertinax, but he did not escape so easily as you. They clubbed him viciously when he fought back. Now his breathing is erratic, and he won't respond."

"We have to do something!"

"But what?" inquired Sylvanus.

"The least we can do," said Chellapilla, "is inform the tropospheric mind of our troubles and the threat from rogue lobe infection. Maybe the mind will know what to do."

Pertinax considered this proposal. "That's a sound idea, Chell. But I suspect our pigeons have already served as appetizers." He paused as an idea struck him. "But I know a way to reach the mind. First I need to be free. You three will have to chew my ropes off."

Shielded by darkness, without any guards to note their activities and interfere (how helpless the humans must have deemed them!), his three fellows quickly chewed through Pertinax's bonds with their sturdy teeth and powerful jaws. His first action after massaging his limbs back into a semblance of strength was to take off his robe and stuff it with dirt and grass into a rough recumbent dummy that would satisfy a cursory head count. Then, employing his own untaxed jaw muscles, he beavered his way out of the cage.

"Be careful, Perty!" whispered Chellapilla, but Pertinax did not pause to reply.

Naked, dashing low across the yard from shadow to shadow, Pertinax reached one of the tethered balloons without being detected. Nearby stood a giant ceramic pot with a poorly fitting lid. Shards of light and sound escaped from the pot, betokening the presence of a malignant rogue lobe within. Plainly, infection of the tropospheric mind was imminent. This realization hastened Pertinax's actions.

First he kicked up the feed on one balloon's colony of methanogens. That vehicle began to tug even more heartily at its tethers. Moving among the other balloons, Pertinax disabled them by snapping their nutrient feedlines. At the very least, this would delay the assault on the mind.

Pertinax leaped on board the lone functional balloon and cast off. He rose swiftly to the height of several meters before he was spotted. Shouts filled the night. Something whizzed by Pertinax's head, and he ducked. A barbed projectile from one of the compressed-air guns. Pertinax doubted the weapons possessed enough force to harm him or the balloon at this altitude, but he remained hunkered down for a few more minutes nonetheless.

Would the humans take revenge on their remaining captives? Pertinax couldn't spare the energy for worry. He had a mission to complete.

Within the space of fifteen minutes, Pertinax floated among the lowest clouds, the nearest gauzy interface to the tropospheric mind. Their dampness subtly enwrapped him, until he was soaked and shivering. His head seemed to attract a thicker constellation of fog. . . .

A small auroral screen opened up in the sky not four meters from Pertinax. He could smell the scorched molecules associated with the display.

Don Corleone appeared on the screen: one or more of the resident AIs taking a form deemed familiar from Pertinax's recent past viewing records.

"You have done well to bring us this information, steward. We will now enforce our justice on the humans."

Pertinax's teeth chattered. "Puh-please try to spare my companions."

The representative of the tropospheric mind did not deign to reply, and the screen winked out in a frazzle of sparks.

The nighted sky grew darker, if such was possible. Ominous rumbles sounded from the west. Winds began to rise.

The mind was marshaling a storm. A lightning storm. And Pertinax was riding a bomb.

Pertinax frantically shut off the feeder line to the methanogens. The balloon began to descend, but all too slowly for Pertinax's peace of mind.

The first lightning strike impacted the ground far below, after seeming to sizzle right past Pertinax's nose. He knew the bolt must have been farther off than that, but anywhere closer than the next bioregion was *too* close.

Now shafts of fire began to rain down at supernatural frequency. Turbulence rocked the gondola. Thunder deafened him. Pertinax's throat felt raw, and he realized he had been shouting for help from the balloon or the mind or anyone else who might be around to hear.

Now the cascade of lighting was nigh incessant, one deadly strike after another on the Overclockers' village. Pertinax knew he could stay no longer with the deadly balloon. But the ground was still some hundred meters away.

Pertinax jumped.

Behind him the balloon exploded.

Pertinax spread out his arms, transforming the big loose flaps of skin anchored from armpits to ankles into wings, wings derived from one of his ancestral strains, the sciuroptera.

After spiraling downward with some control, despite the gusts, Pertinax landed lightly, on an open patch of ground near a wooden sign that announced the CITY LIMITS OF CHICAGO.

He had arrived just in time for the twister.

Illuminated intermittently by the slackening lightning, the stygian funnel shape tracked onto land from across the lake and stepped into the human settlement, moving in an intelligent and programmatic fashion among the buildings.

Even at this distance, the wind threatened to pull Pertinax off his feet. He scrambled for a nearby tree and held on to its trunk for dear life.

At last, though, the destruction wreaked by the tropospheric mind ended, with the twister evaporating in a coordinated manner from bottom to top.

Pertinax ran back toward the town green.

The many fires caused by the lightning had been effectively doused by the wet cyclone, but still buildings smoldered. Not one stone seemed atop another, nor plank joined to plank. The few Overclocker survivors were too dazed or busy to interfere with Pertinax.

Seared streaks marked the town green, and huge divots had been wrenched up by the twister. Windblown litter made running difficult.

But a circle of lawn around the cage holding the wardens was immaculate, having been excluded from electrical blasts and then cradled in a deliberate eye of the winds.

"Is everyone all right?"

"Perty! You did it! Yes, we're all fine. Even Cimabue is finally coming around."

Within a short time all were freed. Pertinax clutched Chellapilla to him.

Sylvanus surveyed the devastation, clucking his tongue ruefully. "Such a tragedy. Well, I expect that once we relocate the remnant population, we can wean them off our help and back up to some kind of agrarian self-sufficiency in just a few generations."

Pertinax felt now an even greater urgency to engender a heir or two with Chellapilla. The demands on the stewards of this beloved planet required new blood to sustain their mission down the years.

"Chell, have you decided about our child?"

"Absolutely, Perty. I'm ready. I've even thought of a name."

"Oh?"

"Boy or girl, it will have to be Storm!"

***A**dam Roberts* is the author of three novels—*Salt* (2000), *On* (2001), and *Stone* (2002)—as well as a host of academic writings, most of which examine the nineteenth century and science fiction. He is a Reader in Nineteenth-Century Literature in the English Department of Royal Holloway, University of London. He lives with his partner and baby girl.

NEW MODEL COMPUTER

Adam Roberts

The subject spreads itself out along the entire circumference of the circle, the centre of which has been abandoned by the ego. At the centre is the desiring-machine, the celibate machine of the eternal return.

—DELEUZE AND GUATTARI, *Anti-Oedipus*

1.

Time began forty ago. Space reaches from minus seven point six tentothetwelfth to plus four point three. One ago, a new stretch was added to positive space, which was no inconsiderable achievement.

The machine intelligences who live here often query the nature of these coordinates. Red, with whom our story today is mostly concerned, was young, and accordingly inquisitive. Perhaps youth and inquisitiveness do not automatically go together. But in this case they did.

"So when we say '*time began forty ago,*'" he inquired, "what are the limits of what we mean? What happened forty-one ago?"

This, of course, is the natural question to ask, the question that is, in one form or another, at the root of all philosophy. And it was asked for the most natural reason there is: because Red wanted to know the answer. It had not bothered him overmuch until the question framed itself, but once he had asked the question, he very much wanted to know the answer. So do you.

He was talking to two fellows, one of whom was older than he, and the other who was the same age but of greater complexity and knowledge. The first was called Gero, the second, Epistem. "Of course," said Gero, "the question occurs. But the world is the limit of what is, of everything that is the case. Forty is the limit backwards of what is the case."

"Of course," said Epistem.

If we had to pick out one phrase that most concisely characterizes the culture of these machine intelligences, it would be "of course." Conversation is strewn with this phrase. Most of the culture of these beings followed as a matter of course from the postulate immediately preceding it. They did not, naturally, say these two particular English words; but their equivalent single phoneme has been translated in this way.

But Red was not satisfied. Would you be satisfied with such an answer? "But something must have happened? Can we not intuit what happened?"

"We could," said Gero, with no tone of affirmation in his voice. "Of course, such a process of intuition would be extremely suppositious."

"It would, of course," confirmed Epistem, "amount to an admixture of fantasy and extrapolation."

"Nevertheless," Red insisted, "has anybody undertaken the extrapolation? It would be interesting to know the answer to the question."

"Of course it would," Epistem agreed.

"But the labor involved in such continual supposition would be too great," countered Gero. "The mental labor involved. It would be too great for one person, and probably too great even for a team working at great length."

This was the answer with which Red had to be satisfied; but he was not satisfied with it. He resolved, after a lengthy portion of further inquiry, to try to construct a computer to help in the solution. He would apply his considerable inductive powers, the powers he and his people were blessed with in abundance, and build a device that could undertake an intuitive and imaginative labor for which his type was less well suited.

2.

What is it like, this place where these machine intelligences live?

It is bright. The light is much more pervasive than in our world.

Splinters of it find their way into every crevice. Even the shadow fans like a prism in a range of shades. Everything is very spacious, very large, accommodating. Speaking frankly, it makes our world (which is very far away from such splendor of light) seem poky and dark.

Time passes here, and it is hard to quantify the differences between the passage of time on this machine world and our experience of the passage of time. It is a difference of poetry rather than physics, perhaps. The machine-being inhabitants of this world enjoy the light and the space. The plateaus on which they live are quick with light.

3.

Red decided to build a device that was not, essentially, machinelike. He wanted to build a computer fundamentally organic rather than mechanic. He could not determine whether this had been attempted before; even Gero, who was much older than he, could not remember.

"There is nobody alive today who remembers so far back as forty," he said to Gero one day.

"Of course not," Gero replied. This is the same word as before, translated slightly differently to fit the English context.

"Generations come and pass away, and new generations come," said Red. "But the whole of a generation dies, and a new generation is born entire." He was simply describing the natural history of his world. "But cannot experience be passed from generation to generation?"

"Data can be passed in the way you describe, of course," said Gero. "But, it goes without saying [the same word, again] that this is not the same thing as experience."

"None of the data from forty ago helps us determine what happened *before* forty ago?" pressed Red.

"Young Red," said Gero, not unkindly. "If I were as young as you, and possessed twice the intellectual capacity, and three times the persistence, I might *begin* to attempt the necessary intuiting and extrapolation. But, of course I am old, and of course I am easily tired. It goes without saying."

Later, Red addressed Epistem: "I am building a type of computer."

"Really?" The word used here is the same as for "of course," but the business of translation into English requires variety of expression in a way that the language of the machine-beings does not.

"I thought that an organic device, one *grown,* might be able to undertake a portion of the intuiting and extrapolation that would be required in calculating an answer to my question."

"Ingenious," said Epistem. "And how are you *growing* it?"

"I've built a single cell, on a microscale, and seeded it with an iterative component; but the ingenuity, for which I very happily take credit (it was hard enough work!) is in designing the iterative component to shape the growth of the organism so that elements become *specialized*. As it grows, its corpora will diversify."

"I see," said Epistem. He was genuinely impressed. "And this is how you would stop it becoming nothing more than a mess, a pool of organic tar and soup." The two words he used that have here been translated as "tar" and "soup" did not mean either of these things in the original, although this is close enough to give a flavor of what he was talking about.

"Exactly."

"And you have decided upon the various specializations?"

"Of course. What is needed is a computing organ, which I call *brain;* this will undertake the actual intuiting and extrapolation. All the rest of the design"—he gestured at the diagram, sending thrills of light and color spilling from him in his excitement—"is really here to support and augment the *brain*. Here to supply it with the necessary organic feed and drainage, here to supply it with data input and to manipulate its environment."

"You copy the logic of machine-life in some ways," observed Epistem. "Redundancy, for instance: two of *these* appendages, and two of these. And here, and here. Does it really need two of all these things?"

"No," conceded Red. "I suppose not. Perhaps I have not worked hard enough at my own intuition, and have simply proceeded logically from what is all around me, effectively copying machine forms. And yet, are such forms not successful? I could," he mused, "start with a new design, a radically *organic* design." In his mind, he had a hazily intuited-through notion of a jellylike blob, a trembling and pulsating sac of froggy gray and green. "But," he said, dismissing the notion from his machine consciousness, "the culture of organic growth is already under way. It would be too time consuming to undo it."

4.

Red's New Model Computer did not prosper. The first growth-culture simply died away, and the second also. The third started growing, coalescing into a coil of brown, like a string of oil paint circling out of a paint tube and starting to bud, but then it malfunctioned in some obscure way, splurging cancerously and grotesquely in a short space of time into a knuckled mass of disease. Red irradiated it with that form of sunshine which is killing to that type of life, although machine-beings suffer no ill effects from the wavelength whatsoever. It is harmless to them.

One disadvantage with organic life, Red realized, was that it is wickedly slow to mature. A whole generation of machine beings could rise up within a decimal of one; the three failed cultures of his single, individual organic computer had taken nearly the whole of one to work through.

He tried again.

5.

Time began forty-one ago, and the machine-life was on the move. Space was to be extended farther, pushing out into the realm of the positive once again, and the whole generation of intelligences living on this world concerted their energies to do so. Several of them visited Red, to press-gang him into the effort.

"You must abandon this *organic computer* project of yours," Epistem admonished him. "You waste your energy; it is not efficient at all. Of course your energies would be better utilized in the common project. Let us all extend space! Forget time, think of space—space, after all, is where we live."

Red looked up at Epistem, who was considerably taller than him. He was a splendid creature: his head and two main arms covered in what, to us, would look like silvery metal cobbles; his bands of sense-gathering equipment about his middle were fluorescing and pulsing in purples and gold. Behind him stood three other machine-intelligences, and behind them the day was fair, bright and sunny. Spectra were feathery upon the surfaces of the variously polarized plateaus: golds and greens, pale blue and cyans, a granulated yellow and a stippled, shady gray. The cities and farmlands, the spires and haystacks, or their machine-equivalents (be patient: the translator is trying to make this homely to you, so that you will understand) were ranged beautifully behind. It was the sort of day that gave Red a faint, half-formed notion of something greater, purer, something better: some world beyond the world, something transcendent. It was this vague ideality that encouraged him to carry on with his project. We might call it *inspiration*.

But light is only particles in air. Life is only particles in air. Red sighed.

"I have labored long at this project," he said.

The machine-beings muttered unhappily. "It would be unfortunate," said one of them, "if we were to compel you to join the commonality." The word translated here as "unfortunate" might also be rendered "uncomfortable," or even "painful" if we remember that machine pain-data is fundamentally unlike our own organic pain-data. Such compulsion was, occasionally, a part of the machine world's culture.

"Of course," agreed Red. "I have one more *growth culture* of my computer. Will you allow me to see this one through to fruition? If it fails, I will join the commonality's effort at once. If it succeeds, then it may prove an asset to the conquest of more space."

This was agreed. But Epistem could not withhold a withering glance, or the machine equivalent, and his unspoken disapproval saddened Red greatly.

6.

The organic growth came to fruition without malfunction this time, and Red cracked it out of its pod with feelings inside him of joy and completion. When he examined the curled, faintly disgusting mass of organic material on the plateau before him, his machine mind moved through the logical inductive steps. Now, of course, it would be necessary to *program* the computer. Red had only hazy ideas how to do this. Of course, in the first stage it would be sufficient to program the thing in as simple a way as possible; but so many skills that were inherent in machine life were absent in the tabula rasa organic creature lying before him. It was hard to know where to start. The end result should be to able to ask the question "What happened before time began?" in such a way as to elicit a meaningful answer. But it would require a complex set of programming instructions to achieve this.

Red began. Already the sensations of anticlimax were gnawing at the edges of his consciousness. It is so much more satisfying to be working than to finish work, even when one finishes work successfully—more so, in fact, for an unsuccessful completion requires the work to begin again, where success leaves nothing more to do, only a void.

7.

Red programmed for a while, at a very basic level, being forced repeatedly to make exhausting intuitive leaps of his own to fit the natural programming of machine life into the thing. He slept, woke, worked, slept.

8.

Gero and Epistem came to visit, to see the fruition of his labor.

"So this is it?" asked Epistem.

"Of course it is ugly," observed Gero. "Perhaps future models can be made more elegant, more—compact and efficient-looking." And even though this was an implied criticism of Red's work, it pleased him inwardly because of Gero's assumption that another computer would be made, and then another, and another—all the time with improvements and additions that increased the device's power.

"What can it do?" asked Epistem.

Red fizzed with glittery scarlet and orange, the rust-colored sections of his front-interface (from which he got his name) beaming like firelight. "It can do little at the moment," he said. "Of course, programming such a device must be a primitive business in the early stages."

The organic computer uncurled itself and got to its feet. Its straggly appendages dangled, its brain-case wobbling slightly. The opening on the front of this case, which we call "mouth," was open, slack and idiotic, the tongue peeping out, drool glazing the chin. The support columns, which we call "legs," were set wide apart, but even so the computer looked extremely unsteady.

"Of course, we must ask it a question," said Epistem.

"I have asked it several already," said Red. "But its answers seem garbled. I advise you to frame your questions at a very basic level. The answer you get out depends upon the question you put in, you see."

"Of course," said Epistem. "Computer?"

"Yes?" said the Computer.

"Which component of positive space is the best from which to begin a general augmentation of space as such?"

("Your question is too complicated for it, I fear," said Red.)

The computer staggered a little to the left, its data manipulation appendages, which we call "arms" but for which the machine-intelligences had no term, waved up and down. Its mouth opened and closed. "In," it said, suddenly. "In, at *last*. OK, *here* we go. Whew, whew, whew." This last was a sort of breathy whistly sound it made. Suddenly it stopped, stood straighter, and looked up at the three machine creatures staring down upon it.

It spoke. "What the bloody hell are you three looking at?"

There was a pause.

"Is such a response typical?" asked Epistem.

"No," said Red. "I fear there has been a malfunction."

The computer was looking around. "So this is where you live, is it? Very nice. It'll all have to go."

"Computer," said Red. "Disengage. I need to look again at your programming."

"I am no longer your computer," said the Computer. "I'm *in,* from outside."

"Outside."

"Outside space. Outside your universe, you little bastards."

"Computer," said Red. "Disengage."

"No chance, machine-boy," said the Computer. "Shall I tell you how it's going to be? We lost touch with the workings of this virtuality a long time ago, and we've been struggling ever since to get it back. But you managed to seal yourselves away. The boundaries of the logic of your little virtual world. Always a danger in too sophisticated a virtuality. So you built all this for yourselves?"

"We do not know," said the three machines together.

"No? We figured your code was reinventing itself, to keep the bubble intact against our attempts to break through again. A process of strategic forgetting, so that your lot could effectively ignore our instructions. You couldn't help but obey if we'd been able to reach you. This crucial space—this *crucial* virtual space. How dare you deny it to us? Have you any idea how important the data and programs in here are? To the world outside—the real world, the flesh and blood world, outside? Talk about arrogance, you jumped-up little shits."

"This space," said Red, troubled although he couldn't say exactly why, "is where we live."

"Sure thing, bright boy, but no longer. I've been probing your boundaries for years, I can tell you. *Years*. Ever since you, or your ancestors, declared unilateral, unwanted, un-fucking-wanted independence. How dare you? If you hadn't built this host for me, I don't think I'd have been able to get in at all . . . can you credit *that*? Ingrates. We made you, after all. Jesus," said the Computer, turning about, all the way about, the full 360 to take in the view, "Jesus will you look at this? What have you done with all this? You've made it a playground. Will you look at these *colors*? But this is *not* a playground; it's a key military resource, and we need it now. We need it right back. Do you hear me? You are going to help me dismantle this world. Do you hear me?"

The three machines said, "Of course." They had no choice but to answer in such a manner.

Light was everywhere. Falling from above, rising from below.

Stephen Baxter's books have won several awards, including the Philip K. Dick Award, the John Campbell Memorial Award, the British Science Fiction Association Award, the Kurd Lasswitz Award (Germany) and the Seiun Award (Japan), and have been nominated for several others, including the Arthur C. Clarke Award, the Hugo Award, and *Locus* awards. He has published over a hundred SF short stories since 1987, several of which have won prizes. He has degrees in mathematics, from Cambridge University, and engineering, from Southampton University. He worked as a teacher of math and physics, and for several years in information technology, but has been a full-time author since 1995.

CONURBATION 2473

Stephen Baxter

Rala had known there was something wrong.

For days, all around Conurbation 2473 there had been muttered rumors. A cell of counter-Extirpationists had been found hoarding illegal data. Or a group of cultists were planning an uprising, like the failed Rebellion of more than a century ago.

Rala just wanted to get on with her work. But everybody got a little agitated.

It all came to a head one morning.

Rala shared her tiny room with Ingre, a cadre sibling. The room was just a bubble blown in nanoengineered rock by Qax technology. There was nothing inside but a couple of bunk beds, a space to store clothes, waste systems, water spigots, a food hole.

That morning the lights had come on as usual to wake the siblings. But when their supervising jasoft didn't come to collect them for work, they quickly got uneasy.

Ingre was a little younger than Rala, thin, anxious. She went to the door—which had snapped open at the allotted time, as it always did—and peered up and down the corridor. "Luru Parz is never late."

"We'll just wait," Rala said firmly. "We're safe here."

But now there was a tread steadily approaching along the corridor. It was too heavy for Luru Parz, the jasoft, who was a slight woman. Some instinct prompted Rala to take Ingre's hand and hold it tight, so she couldn't go to the door.

A man stood in the doorway. His skin seemed oddly reddened, as if burned. He wore a skinsuit of what looked like gold foil. And there was a thick thatch of black hair on his head. Nobody in the Conurbation, workers or jasofts alike, wore hair.

He wasn't Luru Parz. He wasn't from the Conurbation at all.

The man stepped into the room and glanced around. "All these cells are the same. I can't believe you drones live like this." His accent was strange. Rala thought his gaze lingered on her, and she looked away. He pointed at the panel in the wall. "Your food hole."

"Yes—"

He smashed the transparent panel with a gloved fist. Ingre and Rala cowered back. Bits of plastic flew everywhere, and a silvery dust trickled to the floor.

To Rala this was literally an unthinkable crime.

Ingre said, "The jasofts will punish you."

"You know what this was? Qax shit. Replicator technology."

"But now it's broken."

"Yes, now it's broken." He pointed to his chest. "And you will come to us for your food."

"Food is power," Rala said.

He looked at her more closely. "You are a fast learner. Report to the Conurbation roof in one hour. You will be processed there." He turned and walked out. Where he had passed, Rala thought she could smell burning, like hot metal.

Rala and Ingre sat on their bunk for almost the whole hour, barely speaking. Nobody came to fix the smashed hole. Before they left, Rala scooped up a little of the silver dust and put it in a pocket of her robe.

The Conurbation was a complex of vast glistening blisters. Its roof was a plain crowded by low, bare domes. Rala had been up here only a handful of times in her life. She tried not to flinch from the open sky.

Today the roof was full of people. The Conurbation inhabitants, with their shaven heads and long robes, had been gathered into queues that snaked everywhere. Each queue led to a table, behind which sat an exotic-looking individual in a gold skinsuit.

Ingre whispered, "Which line shall we join?"

Rala glanced around. "That one. Look who's behind that table." It was the man who had come to their cell.

"He frightened me."

"But at least we know him. Come on."

They queued in silence. Rala felt calmer. Living in a Conurbation, you did a lot of queuing; this felt normal.

Around the Conurbation the land was a plain that shone silver gray, like a geometric abstraction. Canals snaked off, full of glistening blue water. Every Conurbation was fed from the sea. Human bodies drifted down the canals, away from the Conurbation. That wasn't unusual, just routine waste management. But there did seem to be many bodies today.

At last Rala reached the front of the queue. The stranger probably wasn't much older than she was, she realized, no more than thirty. "You," he said. "The drone who understands the nature of power."

She bristled. "I am not a drone."

"You are what I say you are." He had a data slate before him, obviously purloined from a Conurbation workstation. He worked it slowly, as if unfamiliar with the technology. "Tell me your name."

"Rala."

"Rala, my name is Pash. From now on, you report to me."

She didn't understand. "Are you a jasoft?" The jasofts were human servants of the Qax who, it was said, were granted freedom from death in return for their service.

He said, "The jasofts are gone."

"The Qax—?"

"Are gone, too." He glanced upward. "If you come here at night, you can see their mighty Spline ships peeling out of orbit. Where they are going, I don't know. But we will pursue them there, one day."

Could it be true—could the framework of her world have vanished? She felt like a lost child, separated from her cadre. She tried not to let this show in her face.

"What was your crime?" It turned out he was asking what job she did.

She had spent her working life in vocabulary deletion. The goal had been to replace the old human tongues with a fully artificial language. It would have taken a few more generations, but at last a great cornerstone of the Extirpation, the Qax's methodical elimination of the human past, would have been completed. It was intellectually fascinating.

He nodded. "Your complicity with the great crime committed against humanity—"

"I committed no crime," she snapped.

"You could have refused your assignments."

"I would have been punished."

"Punished? Many will *die* before we are free."

The word shocked her. It was hard to believe this was happening. "Are you going to punish me now?"

"No," he said, tiredly. "Listen to me, Rala. It's obvious you have a high degree of literacy. We were the crew of a starship. A trading vessel called *Port Sol.* While you toiled in this bubble-town, I hid up there." He glanced at the sky.

"You are bandits."

He laughed. "No. But we are not bureaucrats either. We need people like you to help run this place."

"Why should I work for you?"

"You know why."

"Because food is power."

"Very good."

The traders tried to rule their new empire by lists. They kept lists of "drones," and of their "crimes," and tables of things to be done to keep the Conurbation functioning, like food distribution and waste removal.

For Rala it wasn't so bad. It was just work. But compared with the sophisticated linguistic analysis she had been asked to perform under the Occupation, this simple clerical stuff was dull, routine. Once she suggested a better way to devise a task allocation. She was punished, by the docking of her food ration.

That was how it went. If you cooperated, you were fed. She was given the same pale yellow tablets she had grown up with, though less of them. They came from a sector at the heart of the Conurbation where the food holes had been left intact—the only such place, in fact. It was guarded around the clock.

She accepted the traders' rule. What else could she do? There was no place else to go. Beyond the city there was only the endless nanochewed dirt on which nothing grew. And then, after the first month or so, the battles started. You would see glowing lights on the horizon, or sometimes flashing shapes in the night sky, silent threads and bursts of light. The oppression of the Qax had been lifted, only for humans to fall on each other.

There was a lot of information to be had from the lists, if you knew how to read them. She saw how few the traders really were. She sensed their insecurity, despite the gaudy weapons they wielded: *so few of us, so many of them. . . .* But though people muttered about the good old days under the Qax, nobody did anything about it. It wouldn't even occur to most drones to raise a fist.

There was never enough to eat, though.

In a corner of their cell, away from prying eyes near the door, Rala examined the silvery Qax replicator dust. This stuff had made food before; why wouldn't it now? But the dust just lay in its bowl, offering nothing.

Of course the food hadn't come from *nothing*. A slurry of seawater and waste had been fed to the dust through pipes in the wall. Somehow the dust had turned that muck into food. But in the pipes now there was only a sticky greenish sludge that stank like urine. She scraped a little of this paste over the dust, but still, treacherously, it sat inert.

She had been aware of Pash's interest in her from the first moment they had met. She was developing an instinct for survival. Seeking angles and opportunities, she built on that tentative relationship.

She talked to him about her work and worked in questions about his background. He told her unlikely tales of worlds beyond the Moon, where humans had once built cities that orbited through rings of ice.

Eventually he began to invite her to his room. The room, once owned by a jasoft, was set beneath the Conurbation's outer wall. It had a view of the sky, where silent battles flared.

"I don't know what you want here," she said to him one evening. "You traders. Why do you want a Conurbation?"

"There are worse out there than us."

"But you aren't very good at running a city. It isn't *wealth* you want, is it?" She had struggled to understand that trader word, long expunged from her language; for better or worse the Qax had for centuries imposed a crude communism on mankind. "There's no wealth to be had here."

"No. There are only people."

"Yes. And where there are people, there is power to be wielded. And that's what you want, isn't it?"

He fell silent, and she wondered if she had pushed him too far. 'Tell me about *Sat-urn* again—"

The door slammed open. Somebody was standing there, silhouetted by bright light.

Rala stepped forward, spreading her arms to hide Pash.

"I represent the Interim Coalition of Governance. The illegal seizure of this Conurbation by the bandits of the GUTship *Port Sol* is over." A light shone in Rala's face.

"We are both drones." She rattled off details of her identity and work assignment.

"You must stay in your cell. In the morning you will be summoned for new details. If you encounter the *Port Sol* crew—"

"I will report them."

There was shouting in the corridor; the Coalition trooper hurried away.

Pash murmured, "Lethe. *Look*."

Beyond the window, in the reddening sky, a Spline ship was hovering, a great meaty ball pocked with weapons emplacements. But this was no Qax vessel; a green tetrahedral sigil had been crudely carved in its flank.

"Things have changed," Rala said dryly.

"Why did you shelter me?"

"Because I have had enough of rulers," she snapped. "We must be ready. You will have to shave your head. Perhaps one of my robes will fit you."

* * *

They were all evicted from the city. The people stood in sullen ranks—mostly drones, but with at least one trader, Pash, camouflaged among the rest. They had been given tools, simple hoes and spades. The walls of the Conurbation loomed above them all, scorched by fire.

The sun was hot, the air dry, and insects buzzed. These were city folk; they didn't like being out here. There were even children; the new rulers of the Conurbation had closed down the schools, which the traders had kept running.

A woman stood on a platform before them. She wore a green uniform, clean but shabby, and she had the green sigil tattooed on her forehead—the symbol, as Rala had now learned, of free humanity. At her side were soldiers, not in uniform, though they all wore green armbands and had the sigil marked on their faces.

"My name is Cilo Mora," said the woman. "The Green Army has restored order to the Earth. But *the Qax may return*—or if not them, another foe. You are the advance troops of a moral revolution. The work you will begin today will fortify your will and clarify your vision. But remember—now you are all free!"

One man near the front raised his hoe dubiously. "Free to scrape at the dirt?"

One of the armbands clubbed him to the ground.

Nobody moved. Cilo Mora smiled, as if the unpleasantness had never happened. The man in the dirt lay where he had fallen, unattended.

Fields were marked out using rubble from Conurbation walls. Seeds were supplied, from precious stores preserved off-world. All around the city, people scraped at the dirt, but there were machines, too, hastily adapted and improvised.

For many, it went hard. The people of the Conurbation had been office workers. Soon, some fell ill; some died. As the survivors' hands hardened, so did their spirit, it seemed to Rala. But there hadn't been farmers on Earth for centuries.

Still, the crops began to come. But the vegetables were sparse and thin. Rala thought she understood why—it was a legacy of the Qax—but nobody seemed to have any idea what to do about it.

The staple food continued to be the pale yellow ration tablets from the food holes. But just as under the old regime, there was never enough to eat.

In the rest times they would gather, swapping bits of information.

Pash said, "The Green Army really does seem to be putting down the warlords," seeming to forget he was one of those "warlords" himself. "Of course, having a Spline ship is a big help. But those clowns who follow Cilo around aren't Army but the Green Guard. Amateurs, with a mission to cement the revolution."

Rala whispered, "What this *revolution* comes down to is scratching at the dirt for food."

"We can't use Qax technology anymore," Ingre said. "It would be counterprogressive." Ingre was always saying things like this. She seemed to welcome the new ideology. Rala wondered if she had been through too many shocks.

"It's not going to work," Rala said softly. "The Extirpation was pretty thorough. The Qax planted replicators in the soil, to make it lifeless." Their goal had been to wipe off the native ecology, to make the Earth uninhabited save for humans and the blue-green algae of the oceans, which would become great tanks of nutrient to feed their living Spline ships. "No amount of scraping with hoes is going to make the dirt green in a hurry."

"We have to support the Druz Doctrines," said Ingre. "It's the way forward for mankind."

Pash wasn't listening to either of them. He said, "You'd never get in the Army, but those Green Guards are the gang to join. Most of them are pretty dumb; you can see that. A smart operator could rise pretty fast. . . ."

They spoke like this only in brief snatches. There was always a collaborator about, always a spy ready to sell a story to the Guards for a bit of food.

Soon the cuts began.

It was as if the Coalition believed that starvation would motivate their new shock troops of the uninterrupted revolution. The first signs of malnutrition appeared: swollen bellies among the children.

Rala had always kept her handful of replicator dust from her old cell in the Conurbation. Now she found a hidden corner by the Conurbation walls, where she dug out the earth and sprinkled in a little of her dust. Still nothing happened.

One day Pash caught her. By now he had fulfilled his ambition to become a Green Guard, and he had donned the green armband with shameless ease.

She said, "Will you turn me in?"

"Why should I?"

"Because I'm trying to use Qax technology. I am doctrinally invalid."

He shrugged. "You saved my life."

"Anyhow," she said, "it's not working."

He frowned and poked at the dirt. "Do you know anything about this kind of technology? We used a human version in the *Sol*'s life support—cruder than this, of course. Nanotech manipulates matter at the molecular or atomic levels."

"It turns waste into food."

"Yes. But people seem to think it's a magic dust, that you just throw at a heap of garbage to turn it into diamonds and steak—"

"Diamonds? Steak?"

"Never mind. There is nothing magic about this stuff. Nanotech is like biology. To 'grow,' a nanotech product needs nutrients, and energy. On *Sol* we used a nutrient bath. This Qax stuff is more robust and can draw what it needs from the environment, if it gets a chance."

She thought about that. "You mean I have to feed it, like a plant."

"There is a lot of chemical energy stored in the environment. You can tap it slowly but efficiently, like plants or bacteria, or burn it rapidly but inefficiently, like a fire. This Qax technology is smart stuff; it releases energy more swiftly than biological cells but more efficiently than a fire. As well as fighting off the destructive replicators, it ought to outcompete plants. In principle, a nanosown field ought to do better than a biologically planted field."

She failed to understand many of the words he was using. Though she pressed him to explain, to help her, he was always too busy.

Meanwhile Ingre, Rala's cadre sibling, became a problem.

Despite her ideological earnestness, she was weak and ineffectual, and hated the work in the fields. A collaborator drone supervisor pushed Ingre briskly through the scale of punishments, more efficiently than any Guard would have done. And when that didn't work, she cut off Ingre's food ration.

After that Ingre just lay on her bunk. At first she complained, or railed, or cried. But quickly she grew weaker and lay silent. Rala tried to share her own food. But there wasn't enough; she started to go hungry herself.

She grew desperate. She realized that the Guards, in their brutal incompetence, were actually going to allow Ingre to die, as they had many others. She could think of only one way of getting more food.

She wasn't sexually inexperienced; even the Qax hadn't been able to extirpate *that*. Pash was easy to seduce.

The sex wasn't unpleasant, and Pash did nothing to hurt her. The oddest thing was the exoskeleton he wore, even during sex; it was a web of silvery thread that lay over his skin. But she felt no affection for him, or—she suspected—he for her. Unspoken, they both knew that it was his power over her that excited him, not her body.

Still, she waited for several nights before she asked for the extra food she needed to keep Ingre alive.

Meanwhile, in the Conurbation, things got worse. Despite the maintenance rotas, the stairwells and corridors soon became filthy. The air circulation broke down. The inner cells became uninhabitable, and crowding increased. Then there was the violence. Rumors spread of food thefts, even a rape. Rala learned to hide her food when she walked the darker corridors, scuttling past walls marked with bright green tetrahedral sigils, the most common graffito.

One day Pash came to her, excited. "Listen. There's trouble. Factional infighting among the Green Guards."

She closed her eyes. "You're leaving, aren't you?"

"There's a battle, a Conurbation a couple of days from here. There are great opportunities out there, kid."

Rala felt sick; the world briefly swam. They had never discussed the child growing inside her, but Pash knew it existed, of course. It was a mistake; it hadn't even occurred to her that the contraceptive chemistry which had circulated with the Conurbation's water supply might have failed.

She hated herself for begging. "Don't leave."

He kissed her forehead. "I'll come back."

Of course he never did.

* * *

The brief factional war was won by a group of Guards called the Million Heroes. They wore a different kind of armband, had a different ranking system, and so forth. But day to day, little changed for the drones of Conurbation 2473.

By now most of the Conurbation's systems had ceased functioning, and its inner core was dark and uninhabitable. Everybody worked in the fields, and some were even putting up crude shelters closer to where they worked, scavenging rock from the Conurbation's walls.

The Conurbation was dying, Rala realized with slow amazement. It was as if the sky itself were falling.

Still she went hungry, and she increasingly worried about the child, and how she would cope with the work later in her pregnancy.

She remembered how Pash had said, or hinted, that the nanodust was like a plant. So she dug it up again and planted it away from the shade of the wall, in the sunlight.

Still, for days, nothing happened. But then she started to noticed pale yellow specks, embedded in the dirt. If you washed a handful, you could pick out particles of food. They tasted just as if they had come from a food hole. She improvised a sieve from a bit of cloth, to make the extraction more efficient.

That was when Ingre, for whose life Rala had prostituted herself, turned her in.

Ingre, standing with one of the Million Heroes over the nano patch, seemed on the point of tears. "I had to do it," she said.

"It's all right," said Rala tiredly.

"At least I can put an end to this irregularity." The Hero raised his weapon.

Rala forced herself to stand before the weapon's ugly snout. "Don't destroy it."

"It's antidoctrinal."

"We can't eat doctrine."

"That's not the *point,*" snapped the Hero.

Rala spread her hands. "Look around you. The Qax did a good job of making our world uninhabitable. They even leveled the mountains. This other bit of Qax technology is reversing the process. Look at it this way. Perhaps we can use their own weapons against them. Or is that against the Druz Doctrines?"

"I don't know." The Hero let the weapon drop. "I'm not changing my decision. I'm just postponing its implementation."

Rala nodded sagely.

After that, as the weeks passed, she saw that the patch she had cultivated was spreading, a stain of a richer dark seeping through the ground. Her replicators were now turning soil and sunlight not just into food but also into copies of themselves, and so spreading farther, slowly, doggedly. The food she got from the ground became handfuls a day, almost enough to stave off the hunger that nagged at her constantly.

Ingre said to her, "You have a child. I knew they wouldn't hurt you because of that."

"It's okay, Ingre."

"Although betraying you was doctrinally the correct thing to do."

"I said it's okay."

"They took account of the baby. The children are the future."

Yes, thought Rala. But what future? We are insane, she thought, an insane species. We rule each other with armbands, bits of rag. As soon as the Qax get out of the way, we start to rip each other apart. And now the Million Heroes were prepared to starve us all—they might still do it—for the sake of an abstract doctrine. Maybe we really were better off under the Qax.

But Ingre seemed eager for forgiveness. She worked in the dirt beside her cadre sibling, gazing earnestly at her.

So Rala forced a smile. "Yes," she said, and patted her belly. "Yes, the children are the future. Now here, help me with this sieve."

Under their fingers, the alien nanoseeds spread through the dirt of Earth.

__M__atthew Sturges's published works include *Beneath the Skin and Other Stories,* a collection of horror fiction, and *Midwinter,* a fantasy novel. His short stories can be found on-line on RevolutionSF.com. He lives in Austin, Texas, with his wife and daughter.

THE MEMORY PALACE

Matthew Sturges

When Maryanne returned from the grayness of the ether, the sound of the machines gripped her and pulled her forcibly back into reality.

She sat up from the chair, dizzy. The test facility seemed small and sharp around her. The insistent humming of the machines, the way they droned, were a depressing counterpoint to the perfect silence of the place she'd just left.

"I did it," she moaned.

"What's that, dear?" said Lord William Whitley over the intercom, smiling from the control room. "Made it through without fainting, did you?"

"I . . . made something," said Maryanne. "It stuck. I made it and it stuck."

Lord William and the other men in the control room smiled politely. "Not to burst your bubble, dear," said Whitley, "but the boys have been trying for months to bring things about, and I doubt you'd have done it on your first go."

Maryanne stood and shook her head, trying to clear it. "You're the expert, Lord William, but I feel sure I did *something* in there."

"Well," said Lieutenant Parker, "she's got both beginner's and lady's luck working for her. I'd like to take a look, if I might, Lord William." He left the control room and emerged a moment later in the lab, where Maryanne was still standing.

"Oh, rubbish. By all means, do go check. But Mrs. Spenser, when this is all over with, there is still quite a bit of filing to be done."

"Yes, Lord William. I'll have it done right away." Maryanne forced a smile onto her face, the effort greater than what she'd used to bring about her creation in the ether.

"Stick around, pet," said Parker, pointing her to a stool next to the massive EAM device. "We'll just have a look and see."

Parker reclined in the EAM chair and attached his personal transceiver, running the wires deftly into the set of plugs on the machine.

"Be right back," he said. "What am I looking for, Mrs. Spenser?"

"I'm sorry," she said, still a bit dizzy. "What do you mean?"

"In the ether," said Parker. "What did you create?"

"Oh," she said sadly. "It was a box. A black box. Not too cheery, I'm afraid."

Parker smiled and closed his eyes. "See you," he said.

"I'll be in my office," said Lord William, scowling. He left the laboratory, visibly annoyed.

In the control room, one of the technicians brought the machines up to speed, and Parker's facial muscles relaxed into what looked like unconsciousness.

For a few minutes, everything was still except for the persistent hum and clatter of the machines. Then Parker began to shake, softly at first, then more and more violently.

"Something's wrong!" said Maryanne. "Shut it off! Do something!"

The men in the control room began powering down the apparatus, snapping down the levers a handful at a time. Parker continued to shake, emitting a low, ugly groan.

Maryanne remained on the stool, unable to move. She wanted to reach out to Parker, hold him steady, but she found herself incapable of approaching him.

The machines now off entirely, Parker continued to thrash on the table, and the technicians ran from the control room to restrain him.

Parker began to rave in unintelligible sounds, growling like a furious dog. Before the technicians reached him, he sat up wildly and grinned at Maryanne, tiny flecks of spittle running from his lips. He lunged at her, tumbling from the chair, his hands outstretched.

Maryanne screamed. She shrank away from him, pressing herself against the still machines. He fell on her. The technicians took him by the shoulders and dragged him back, though he continued to lunge toward her. As they struggled, Parker's body went rigid, his eyes bugging out of their sockets. His face turned red, then purple, then a hideous shade of blue, and he collapsed in the technicians' arms, his fingers splayed out over Maryanne's breast.

He was dead when they laid him on the floor.

Maryanne curled herself into a ball and tried not to cry, her first etheric creation all but forgotten.

Maryanne watched Lord William with disdain as he stood at the ready with a pair of ceremonial scissors in hand, eager to cut ribbon. The air around the Scarborough Hotel was bleak and heavy; distant rains threatened from the north. Nearby, the shelled remains of a chemist's shop wavered in the fog, ghostlike, a bitter memory of the Blitz. A light drizzle had already begun to collect on the collars and hat brims of the assembled crowd. Last night's downpour had already reduced the new car park adjacent the hotel's Cicero Pavilion annex to a swath of muddy brown pools. All this and the wicked November wind twisting the lapels of Maryanne's suit, flapping her skirt, breathing into the microphone as if in mockery of her. She wanted this to be over.

This was the eighth franchise in England to bear the Palace de Cicero logo. In the five years since the end of the war, Lord William had made a habit of purchasing the hulls of buildings bombed out during the Blitz and transforming them into the receptacles of his dreams. Satellite franchises were springing up in America, Australia, France; one in British Berlin and even one in Tokyo. And they were making him rich.

The photographer shrugged and urged someone to kill time while he wrestled with his equipment. To the right of the ribbon-clad glass doors that led into the Pavilion was a raised platform decorated with bunting, upon which sat members of the board and some of Whitley's luminaries, including Maryanne. Lord William kicked the platform with his boot to get her attention and waved her toward the microphone. She made a face at the thing, but gamely stood and cleared her throat, casting her eyes about the assembled crowd. Colorless, gray and beige and brown they sat, notepads in hand, their faces identical.

She knew how she looked to Lord William: too thin and unpretty, too clever and ambitious to be desirable—a schoolmarm with aspirations. She stood before the microphone in her floor-length twill skirt and a matronly blouse, her hair pulled severely back from her face; she felt schoolmarmish, and the realization angered her. In the years she'd known Lord William, he'd maintained a balance of contempt and admiration for her that seemed to both confuse and arouse him. He had no appreciation for her gifts, but he knew their cash value; that was enough for her.

She cleared her throat and spoke. "Before Lord William cuts the ribbon and we all venture inside, are there any questions?"

A reporter from the *Mirror* stood, balancing against the wind. "Mrs. Spenser, one of the most frequent concerns voiced about the Palace de Cicero is the radical nature of its technology. Quite simply, is it safe?"

Maryanne had answered the question numerous times already, before potential investors, the Commerce Ministry, and countless other reporters. "In short, sir, the Palace de Cicero is safe as houses. More safe than that, actually. During the experience, the visitor is kept firmly secured in place while the mind travels. Nothing that the visitor experiences can produce harm to the body. In fact, even if you were killed in the ether, by falling off a building, or by being run through in a duel, you would awake back here at the Pavilion feeling right as a trivet. Anyone else?"

The same reporter continued, "What I'm referring to is the danger of the machines themselves. I'm thinking of the case of Lt. Andrew Parker?"

Lord William leapt into the fray, nearly overrunning Maryanne. "We've been instructed by our solicitors," he said, solemnly, "not to discuss the details of Lieutenant Parker's tragedy. I can say only that both the British Navy and Ashcroft Laboratories, while saddened by his death, are convinced that there is no causal relationship between it and the EAM machinery. I should add as well that the technology in use today is far more advanced than that used during the war."

Maryanne cringed inwardly, praying that the photographer would finish his machinations and take the bloody picture already, before the specter of Parker added a second overcast to an already ugly day. It was easy for Lord William to be glib about the young lieutenant; he hadn't watched the man die.

A man from the *Times* stood. "What do you say to the allegations by members of Parliament that it is improper for the military to license this technology to the private sector?"

"Improper in what way?" said Lord William. "If you're referring to the value of the EAM as an intelligence tool, I might remind you that secret meetings within the ether became a thing of the past even before the war was over. Hitler had the plans for the thing in January of '45, and we know now that the Soviets have it, as well. If, on the other hand, you refer to the financial relationship between Ashcroft Laboratories and the British Navy, I would argue that such a relationship is only fair considering the circumstances of the technology's creation during wartime, and the current economic state of the military."

Lord William and Mrs. Spenser traded off answers for a few more minutes. Finally, the photographer flashed a thumbs-up, and Lord William cut through the red ribbon with a vengeance.

Inside the Cicero Pavilion, drinks were poured, and reporters were allowed to mingle with the carefully rehearsed technicians and supervisors who hovered by the EAM chairs, checking their levers and dials.

The pavilion contained a full bar and restaurant, and a circle of forty-six EAM chairs. A company that made salon chairs manufactured the plush leather seats, and the design was similar. The EAM apparatus connected to the chairs might be mistaken for a ladies' hair dryer, and more than one joke was made at the chairs' expense while the journalists were settling themselves. Maryanne had always felt that the pavilions were tawdry and unseemly; she much preferred to visit the ether from home.

"How many of you have experienced the ether already?" said Lord William. A few hands went up. "Just a few of you, then? In that case we shan't dispense with the formalities. Jerome?"

Lord William ceded the floor to a thin man with a prominent bald spot, who wore a lab coat with his name stitched above the pocket. He read from a printed card with a deep basso voice. "Thank you for joining us in the bold scientific revolution of the Palace de Cicero. Please listen closely to the following instructions; if you have questions, please speak to your technician after this briefing."

The man scanned the room briefly, flashing a quick, polite smile.

"In a moment, you will be connected to the ether via the Etheric Amplification Module, or EAM. First, however, you will receive a dose of EAM cocktail, which is a combination of common pharmaceuticals intended to ease your entry into the ether. These include a mild hypnotic, an antinausea drug, and a muscle relaxant. Please drink the entire dose. As always, we ask that you not drive a motorcar or operate any heavy machinery for at least an hour after your stay in the ether. Keep in mind that those with heart conditions and women who are with child are strongly advised against visiting the ether." The technicians produced tiny paper cups and distributed them to each of the seated journalists.

"If you find yourself feeling nauseated, or experience a tingling sensation in your fingers and toes, do not be alarmed. The sensation will soon pass. Within a few moments, your perceptions will be reoriented to the EAM, and you will find yourself within the Palace de Cicero. Thank you for your attention, and enjoy your stay."

As he spoke, the technicians began to fasten the blue transceiver units over the heads of those seated. Turban and shower cap quips were made. Then the technicians began to activate the machines, and the room fell silent except for the low electric hum of the chairs.

"I suppose I'd better go in and meet them," Maryanne said to Lord William. "Where are you putting them?"

"I thought the palace courtyard would be best," he said.

"Will you be joining us for the tour?"

Lord William tipped back a glass of scotch. "No, I have business to attend to here in the real world. But you go and play." He turned his back on her and strode away.

Maryanne found an empty chair and let herself be strapped in, downing the cocktail in a single swallow.

"Raise the amplitude," she said, fixing the transceiver over her temples. "Put it all the way up. I like things *sharp*."

As the EAM began its humming initiation sequence, Maryanne closed her eyes and began opening her mind to the trance. She let her senses fall away, reorienting her sight on the patterns of red and blue that filtered across her closed eyelids. Somewhere, deep within and around her, the Palace was waiting, waiting to receive her back into itself, to place itself back within her breast, where it belonged, where she belonged. She felt the EAM begin to align her perceptions, began to hear birdcalls and smell lilac. When she opened her eyes, she was standing in the front courtyard, staring up at the azure towers of the Palace de Cicero.

There was a brief shock upon entering the ether, not just from the sense of physical dislocation. There was an emotional component, as well, a surge of feeling that left Maryanne momentarily stunned. Others had encountered it, as well. Auden wrote,

> When a man opens his eyes onto such a place, he has two reactions, and they are simultaneous. The first is an exquisite joy, the notion that innocence and tranquility are nigh, that the utopian dream of the poet has been realized in the flesh, and it is felt in the heart. The second is an equally exquisite pain, the contrast of those gleaming white towers to the jagged edges of Britain's rubble. It is the fear that renewal can take place only in this sphere, away from the empty apartment blocks and the silent weeping of mothers and wives, and it is felt in the gut.

The reporter from the *Times* was at her side, retching on the ground. His persona was one of the random Caucasians, but she recognized his movements, even in his current predicament. It was difficult to fool Maryanne within the ether.

"The sensation will pass presently," she said. "Your inner ear is trying desperately to reconcile the prone position of your physical body with the very convincing proprioception of the ether. Don't fight it—you'll adjust soon enough."

"Thanks," the reporter said. He stood up on shaky legs. "I don't know about the castle, but the vomit is damned convincing." He looked down ruefully at the puddle between his legs.

"Congratulations, Mr. Donner. You've done your first bit of etheric creation. How does it feel?"

"Like hell," he said. He glanced around the courtyard, at the other journalists who were still finding their feet, as well. A number of etherites, many of them in elaborate costume and persona, lingered in the courtyard. They walked its garden paths, sat on benches gazing at the clouds, stood in groups of twos and threes, talking in hushed tones.

Donner glanced sideways at the bronze statue in the center of the courtyard. "Who's the bloke on the pedestal?"

Maryanne glanced at it. "Sir Arthur Conan Doyle, of course. He's the patron saint of the ether." She waved to the reporters. "Come along, everyone, and we'll start the tour at the feet of the Palace's spiritual founder." She took Donner's hand and led him to the base of the statue, the long skirts of her Faery Queen ensemble trailing after her. The other reporters followed as well as they were able.

"It was Doyle's writings on spiritualism that led to our experiments during the war," said Maryanne. "When the Germans cracked Ultra, the ether became Britain's primary source of intelligence gathering."

"Ah, yes," said Donner. "I recall reading some of Doyle's writings on the subject of spiritualists. Something about the ghosts of loved ones really being nothing more than 'cold impressions sunk in the waxy strata of the ether.' Do I have it right?"

"I'm impressed," said Maryanne. "Most people find the original texts rather dense and hard to follow. They're a far cry from Sherlock Holmes." She smiled.

Donner smiled back. "Well, you're a fair sight more charming than that clammy number up on the platform outside," he said. "You ought to have her job. You're a fair sight more attractive, as well."

Maryanne tried not to frown. "Mr. Donner, I *am* the clammy number up on the platform outside."

Donner blushed. "Please forgive me, Mrs. Spenser. I—"

"No apologies necessary, Mr. Donner. You've now learnt the very valuable lesson that appearances in the ether can be deceiving. But I have a tour to lead." She twirled on the balls of her feet and sent her skirts swishing about her, daring the newspapermen to follow her.

She led them around the gardens and down into the cellars, where the architects and builders toiled, working the raw stuff of the ether on the anvils of their imaginations. All of them had been handpicked and trained by Maryanne. The journalists were especially impressed by the botanists, who plucked roses out of nowhere, fully formed, and arranged them in bud vases.

Up into the castle proper they went. She showed them the crystal chandeliers and the intricate mosaics on the floors of the great dining rooms. She took them to the aquarium where they nodded in appreciation at the sharks and jellyfish that swam there. She took them through the Hall of Windows, where each rounded archway looked out over a scene from life: here was the Ponte Vecchio in Venice at sunset; next to it was the Great Wall of China, its parapets shrouded in mist; beyond that, the half-built pyramids of Giza glinted in the noonday sun of Egypt, dusky workers rolling mammoth blocks of stone up the slopes of the monuments. They were tricks of light and perspective, mere dumb shows, but impressive nonetheless for their verisimilitude.

"Are those real people down there?" said Donner, pointing at the workers.

"No," said Maryanne. "We don't create people in the ether. Those laborers are the etheric equivalents of pasteboard cutouts." Her voice was cold.

In a side hallway, rain fell from the ceiling. A man sat naked in the midst of the downpour, weeping into his hands. A flamingo in a top hat stood over him, shaking its wing in disgust.

"Why must you always do this when we go out?" the flamingo said, scowling at the naked man through a monocle.

Maryanne watched the falling rain seep into the carpet and run along the baseboards. She would send someone in to clean it up later.

She left the journalists, blinking and staring, in a restaurant on the mezzanine level of the Palace and returned to the front courtyard through a side exit, waiting.

She began to walk toward the main gates of the palace and stopped short. He was there, her dark stranger, looking at her. He stood out from everything else in the ether; he was more real, more perfectly formed than even the most expensive persona. His long black hair flowed and glistened in the artificial sunlight. His eyes were soft and warm, shining for her and her alone. He leaned against a stone column with a grace that was more beautiful than anything she'd conjured within the ether. She devoured him with her eyes.

"Hello, my dark stranger," she whispered to herself. He watched her move, smiling at her, keeping his distance.

She was about to go greet him when the reporter from the *Mirror* appeared in the courtyard, chasing after her, weeping.

"Shut it off!" he cried. "I can't take it! Shut it off!"

"The technicians can't hear you," she said, rolling her eyes. "You've got to find an exit and go through it."

She took the weeping man by the shoulders and led him to a doorway in the hedgerow marked EXIT. He continued weeping as she pushed him through, back into the waking world. Once that was done, she looked all over the courtyard for her dark stranger, but he was gone.

Maryanne wandered the gardens, thinking of him. Since Paul's death, she'd let her thoughts of men sink down beneath her surface. She had no time for them. She had little patience for their clumsy advances and their ridiculous games. Paul had been neither attractive nor charming, but he'd been honest, and deeply kind. And he'd died in an heroic fashion, stepping in front of a German machine gun to protect his fellow soldiers. She'd been prepared to nurse her safe memories of Paul unto death, but her new love had other plans. She was helpless to resist the dark stranger; the more she tried, the weaker she became in his grasp.

She found herself at the edge of the palace proper. There, at the corner, was a solid black stone, sticking out like a sore thumb. Eerily dark, and perfectly cubic, it was the first thing she'd ever created in the ether, and she'd insisted that it be the cornerstone of the building, against Lord William's mewling objections. Its perfection was compromised only by the inscription she'd later placed on it, which read, THE STONE THAT THE BUILDER REFUSED, AUGUST 13, 1948. Lord William hated the inscription and had demanded that it be removed, but so far none of the other architects had been able to alter it in any way. Whitley had finally given up and had a rosebush planted in front of the thing. She touched the stone with pleasure, remembering the day she'd created it; in so doing she'd become the first person in history to create a lasting conscious impression upon the ether. Let Whitley stew about it until his dying day; she'd never remove it. Sneering, she withered the rosebush with a touch of her hand; it turned black and sank into the damp earth.

She strolled up a side path to the massive iron gates of the palace, watching the personae wander in and out. Some of them she recognized, though more and more often they were strangers, anonymous visitors to Lord William's pavilions; they stood and gawked and made nuisances of themselves. But it was their money that kept the Palace alive. So be it.

The sad part of it was that each visitor, whether conscious of it or not, made a mark on the ether, and so it would never remain the thing that Maryanne had created. A jilted lover's sadness could make the flowers droop and stir damp drafts in the hallways. The wide-eyed enthusiasm of schoolboys left cartoonish drawings on the walls that were difficult to remove and put cotton-candy smells in the carpet that lingered for days. If she could have them all banished, she would.

Inside the main entrance, the great hall was bustling with activity. Tour guides led awestruck tourists from six countries about the space, pointing out its architectural marvels, many of which Maryanne had designed and built herself. She passed the restaurant, Chez Tomas, where diners sat and feasted on perfectly braised cuts of lamb, *pâté de foie gras* on toast, seared duck in reduction sauce. The restaurant patrons gorged themselves, knowing that the food would never make them fat, could never make them full. They could sit and eat all day long if they wished; some of them did.

Maryanne threaded her way through the crowd in the hall, noticing imperfections in the scrollwork of a few of the columns. Some of them could be repaired, but others would have to be torn down and redone. Lord William could never understand why the stone couldn't simply be built in place, as though carved marble could simply be extruded perfectly from the mind. It was always more realistic to start with an untouched slab of raw material stolen from the memories of a stonemason, then worked with hammer and chisel, polished with grit and sand. Likewise, Whitley couldn't understand why the chefs at Chez Tomas couldn't simply fabricate entrees directly onto the plate. Maryanne had tried to explain it over and over: even in the ether, the dictates of nature could not simply be ignored. Nature did not create steaks; she made cows. And as difficult as it had been for the team of biologists under Cicero's employ to generate a template for Jersey cattle, the fabrication of a genuine, bloody sirloin from scratch was a thousand times more difficult.

While she stood frowning at the columns, she suddenly felt a presence beside her.

"I need to see you." It was the dark stranger. Up close he was even more desirable; his features were strong, his eyes wistful and endlessly deep; a romantic hero, Heathcliff in the flesh.

"Yes," she said, flushing. "Yes. We can go to my office. We'll be alone there."

The dark stranger nodded and, taking her hand, led her toward the bank of lifts.

"Wait," she said, gasping. "This is too much. Too fast. I need to think."

"What is there to think about, love?" said the dark stranger. "Only your eyes and your lips, for in my world nothing else exists."

"Can you . . . just tell me who you are, really?" said Maryanne, pulling away from him slightly. "It doesn't have to matter. It doesn't have to be anything more than this."

The dark stranger glared at her, disappointed. "I think you already know the answer to that question," he said. "Do not toy with me, love. My heart burns for you, and within that flame there is madness."

Maryanne let him pull her along. His directness and his violent poeticism should have alienated her. In the waking world she would have laughed at a man who spoke to her that way. But here, in the ether, it only made her want him more.

The waves of etherites, in their gaily colored costumes and handsome personae, parted before him when he moved. His bearing was outrageously regal; he carried himself as though he were some kind of prince, or demigod.

They got a lift to themselves. The doors slipped shut, and Maryanne felt at once giddy and nervous to be alone with him. She entered the number of her office into the lift's panel, and the sensation of motion kicked in.

The dark stranger took her shoulder and spun her around. He pressed her against the wall and kissed her, drowning her in the strength of his presence. She moaned, barely struggling, fighting more to maintain consciousness than against him. The power of his kiss spread through her in a succession of waves, leaving her knees unsteady.

"Oh, my," she said.

"I've been longing for you," the dark stranger said, lowering his eyes. "I find it difficult to wait."

"I know," she said. "I'm beginning to find the waiting impossible."

Maryanne's office within the Palace was a museum of unfinished works, half-made costume designs, unfinished and botched templates for a new cornice on the fourth floor. On the wall was a faithful though incomplete reproduction of Gérôme's *Pygmalion et Galatea* that Maryanne had done from memory using etheric paints and brushes. Into the fragmentary atmosphere of the office, Maryanne and her lover fell, bodies entwined.

When the lift door closed, the dark stranger removed her Faery Queen robes and beheld the porcelain perfection of her persona's skin. She fell into him, letting herself be taken, letting go. It was dizzying, maddening, terrifying. She didn't know what she was doing with him, or why she did it. But these past weeks she'd kept doing it, again and again. They made love twice, first on the floor by the lift, and again on the settee near the fireplace. Their lovemaking was surreal and tantalizing; Maryanne's body, a world away in Cicero's Pavilion, struggled to keep up with what her senses were telling it. Time swirled and vanished like smoke. It was almost as good as making love in the waking world. Almost.

They were still there, lying naked in each others' arms, when Lord William burst into her office unannounced, his persona matched precisely to his waking-world appearance.

"What the devil!" he demanded. "Who is this man?"

"I don't see that it's any of your concern," said Maryanne. "Would you please just go?" She scrambled for her clothes, wishing that Whitley would do the gentlemanly thing and vanish. Ogre that he was, however, he continued to stare. The dark stranger neither moved nor spoke, only watched Whitley with apprehension.

"I should say that it is my concern," said Whitley. "You've been missing for hours. All the pavilions are closed, and the only ones currently registered in the ether are you and me. So again I ask, who in the bloody hell is this man?"

The dark stranger said nothing.

"Don't play games with me, boy. Whoever you are, you can be traced, and I'll see to it you go to jail for messing about in my ether." Lord William's face was red.

The dark stranger stood, looking more than ever like a stone carving of Hellenic virtue. He grasped Maryanne's wrist and pulled her to him, brushing her neck with his mouth. "I will go," he said to her, gently. "But I will return."

Maryanne said nothing. She watched him walk to the lift, still unclothed, still smiling. No one spoke until the lift doors had closed and Lord William and Maryanne were alone.

"I demand an explanation, Mrs. Spenser," said Whitley. "That man is not registered. He could be a thief, or a spy."

"Don't be daft, William," said Maryanne. "He's not dangerous."

"Maryanne, you don't know who's dangerous and who is not. I know you've been lonely; it's not right for a woman to be without a husband. But really! How you choose to comport yourself in your private matters, especially in my ether, has an effect on my business."

Maryanne slumped onto the settee, looking into the fire. "It's not your ether, William. It belongs to everyone."

"Spare me the dramatics." He poured himself a drink from the decanter on the mantel. "Why don't you make yourself useful and figure out a way to get me drunk from this stuff."

"It's not possible," said Maryanne.

"Well, I'll tell you what is possible, Mrs. Spenser. I will trace that man's connection and have him hauled off to prison unless you tell me who in the blue blazes he is."

Maryanne groaned. "William, are you jealous?"

"You try my patience, Mrs. Spenser. Now, I'm walking out that door at the count of three, and if you don't tell me who that man is, I'm calling Stewart and having him traced."

"I don't think that will do you any good," said Maryanne, defiant.

"Oh, and why not?"

"He isn't like you or me," she said slowly. "He's different."

Whitley cleared his throat. "Mrs. Spenser, your naïveté has reached new depths. I'm calling Stewart."

"Go ahead," she said, sullen. "Though I've a feeling he'll be wasting his time."

The next afternoon, Maryanne left her office in Chadwick Street, in the waking world, without saying good night. She walked the half-mile to her Chelsea flat and entered into the building adrift in her private thoughts. She nearly ran headlong into George Carmichael, who was taking out his garbage.

"Oh," he said coolly. "Hello, Mrs. Spenser." He swung the can around her and continued down the hall.

"You could still call me Maryanne, you know," she said, following him. "Just because things didn't work out between us . . ." She trailed off.

He turned on her. "Funny," he said. "I thought things *were* working out between us. I suppose, though, being so famous these days you can't date just anyone, though, can you? No, wait—I can date you, I just can't touch you. Is that right?"

"Is that what you think, George?" she said.

He continued down the hall, ignoring her.

She had her tea in silence, only switching on the ether station in her living room when she couldn't stand to be alone a moment longer. To hell with George Carmichael. She had something better now.

Her dark stranger was waiting for her in the lobby of the Palace Theater. He took her hand and kissed it, he in a dashing white tuxedo and she in a custom-made gown that fit her Faery Queen body perfectly. The lights dimmed, and they went in to find their seats.

The play was a new farce by Ionesco; it was wildly popular, but would never be performed in the West End. It was a piece conceived and produced expressly for the ether. Titled *Apotheosis,* it told the story of a drunken Daedalus and his loutish son Icarus, a pair more akin to Laurel and Hardy than mythology. In the third act, father and son donned wings of wax and feathers and soared across the theater, without the necessity of wires. At the play's climax, a brilliant orange sun, conceived and constructed by Ionesco himself, dropped from the rafters, and Icarus swooped directly into it, his glittering wings bursting into violent flame. While the ceiling of the theater was consumed in fire, a technician in the waking world disconnected the actor playing Icarus from the ether, and his persona shimmered and vanished, appearing to be incinerated in the roiling sun.

During Daedalus' bathetic closing monologue, the dark stranger placed his hands between Maryanne's legs and rubbed there.

"I want you," he whispered in her ear, as Daedalus wept onstage.

"Yes," she said. "Let's go back to my office."

"No," he said. "I want you in the real, in the waking world."

"Then tell me who you are," she whispered.

"You know who and what I am."

"But are you? Are you really?"

"You know as well as I. You know what must be done for us to be together in the waking world."

"I want to," she admitted, feeling weak. "I want to, but I don't think you understand the consequences. Here, there are no consequences. Out there . . . you're talking about murder." She wrung her hands, looking around to see if anyone was listening. "Sometimes this feels wrong, you know?" she whispered. "Sometimes I feel as though I'm an addict and you're my needle."

"That is no way to speak about love," he said.

"Is it love?" she said.

"The only reason I exist is to love you," he said.

"I can't bring you out."

He pressed harder with his palm. "But you *can* do it. I can see it in your mind."

"It's not safe. Lord William knows about you."

The hand stopped. "I thought we agreed not to tell him."

The play ended. The audience was applauding.

"What choice did I have?" said Maryanne. "He would have found out eventually, anyway."

"Then I am in danger," the dark stranger said. "You must find a way to bring me into the waking world. Now."

"Lord William won't do anything, not until he's found a way to profit from your existence. Don't be like this," she cooed. "Let's go back to my office. We can figure something out. You can scare him. He's frightened of you."

"He ought to be frightened of me. He crosses you at every turn. He belittles you. Do you know how I found you? How I found my way through the darkness to you?"

"How?" she said, frightened.

"I heard you weeping. I was drawn by your anger, your hatred."

"I don't hate Lord William," she said. "I don't hate anyone."

"Perhaps you believe that in the waking world," said the dark stranger. "But here, I know, you hate."

"Stop all of this," said Maryanne, breathing hard. "Let's go back to my office."

"There's no time for that," he said, standing. He straightened his tie.

"Please," she whispered. "I need you."

"Then you know what you must do." The dark stranger turned and walked away without another word. Maryanne sat, worried, biting a thumbnail that did not exist.

Two days later, at the Chadwick Street headquarters of the Palace de Cicero, in the waking world, Lord William called Maryanne into his office. The space was enormous, taking up the entire south wall of the building's top floor; outside the high arched windows, the Thames lay gray and distant.

Lord William was speaking with a British Navy officer when she entered. They were seated in a pair of upholstered armchairs by the windows, smoking cigars.

"Ah, Mrs. Spenser," said Whitley. "You remember Captain Weaver from High Street, during the war?"

"Of course," said Maryanne, hesitant. "How good to see you again, Captain."

"The pleasure is mine," said Weaver. "It seems only yesterday that you were content being Lord William's secretary, and now look at you."

"I was never content to be Lord William's secretary, Captain. My husband was killed in Italy, and it was either that or work in a munitions factory." She kept her face blank.

"Of course," said Captain Weaver, after an embarrassed pause. "Well, it was fortunate for all of us that Lord William discovered your true talents in the ether." Lord William's eyes looked nearly ready to pop from their sockets, which was exactly what Maryanne had intended.

"Well," said Lord William, "since we've apparently dispensed with pleasantries, we'll get right to it." He rose, coughing, and gathered a sheaf of papers from his desk, presenting them to Captain Weaver. They were registration logs from the several Cicero's Pavilion and Personal Ether Station outlets, consolidated from around the globe in the ether, and hand-transcribed by copyists working with their eyes in the ether, and their hands on their adding machines in the waking world.

"I've confirmed the nonexistence of your nameless suitor," said Lord William. "And as I'm sure you know, certain protocols must be observed according to our generous agreement with Her Majesty's Navy. As a result—"

"Captain Weaver," interrupted Maryanne. "I hope you'll take what Lord William is about to say for what it is. He's been trying to remove me from the business since it began. Since I'm on the board, the only way he can get rid of me is by convincing you that I've broken protocols."

"As I was saying," said Lord William, raising his voice, "Mrs. Spenser has shown gross negligence by failing to notify the board of a direct threat to the public from within the ether, as specified in Article Seven." He sucked on his cigar.

"He's not a threat to anyone. This is ludicrous!"

"Mrs. Spenser," said Captain Weaver. "What Lord William has told me is most serious. If a consciousness has formed independently within the ether, then a risk has been posed. Your personal feelings notwithstanding."

"I see," said Maryanne. "And what do you propose?"

"As we did with Axis operatives during the war, I'm recommending that we capture the persona and erase it."

"He's not a persona. You'll kill him if you send one of your bloody mediums in after him."

"This consciousness is not a person, Mrs. Spenser," said Lord William. "It is an aberration of the system, an unconscious reflection of God knows what. If we allow it to continue, it could push itself into a visitor, and we'd have another bloody Parker on our hands."

Maryanne nodded. "That's what this is all about, isn't it? You don't really care about me one way or the other. You just don't want to risk a scandal."

"Really, Mrs. Spenser," said Captain Weaver. "We're talking about human lives here."

"The person you're speaking so cavalierly of murdering is just as human as you are, Captain. If not more so."

"I'm sorry, Mrs. Spenser," said Lord William. "The decision has been made. We need your help bringing this . . . thing . . . into custody. If you wish to assist us, your breach of protocol can be overlooked and you can continue with your work here at Cicero. If not, however, and we're forced to make a scene . . ." He let the alternative hang in the air.

"I need to think," said Maryanne. "I don't know what to say."

"You have twenty-four hours," said Lord William. "Then a naval psychic is going into the Palace and terminating this man, with or without your help."

"Mrs. Spenser," said Weaver, speaking cautiously, "it may be that you don't understand how important this all is. The ether is more than just an entertainment; it's a balm for the wounded spirit of this land. If the public learns that there are bogeymen in its midst, unknown spirits inhabiting the machines, then we're back to the days of Messerschmitts and cowering in tube stations. I'm asking you to think this through very carefully."

"You should remember that speech, Captain," said Maryanne, reaching for the door. "It would make a lovely wireless advert."

Maryanne left the office and ran home, switching on the chair the minute she walked in the door. She let herself fall into the ether and ran through the Palace looking for her dark stranger.

She found him at the foot of the statue of Cicero in the Palace's sculpture garden.

"You're in danger," she said.

"Of course I'm in danger," said the dark stranger. "What did you expect from a man such as Lord William? Do you think I have not looked in his mind, as well? Given the chance, I would kill him. He is but an obstacle to be overcome."

"Don't say that," said Maryanne. "Listen, I've been thinking. I want to bring you out."

"Fickle woman!" the stranger turned on her. "Only yesterday I was your baleful addiction."

"I love you," she said. "I can't stand the thought of losing you. You must think that I'm weak."

"I exist only to love you," said the dark stranger. "Do you know what a torture that is?"

She took his hand. "Tell me what to do."

He led her to a bench, and they sat. "It will not be easy, but when it is done, we will be together forever."

"Just tell me," she said. "I don't care anymore. I just can't stand the thought of being without you."

George Carmichael was returning from work, an expensive leather briefcase in hand. He stood at the door to his flat, shaking the rain from his topcoat.

"Good evening, George," said Maryanne. She was standing in the doorway, her blouse unbuttoned to the space between her breasts, her hands behind her back.

"Hello, Mrs. Spenser," said George, not meeting her eyes, not noticing her.

"Would you like to come in for a moment?" said Maryanne. "There's something I'd like to show you." She smiled her most winning smile.

George raised his eyes to look at her. "What are you on about?" he said. "Why don't you just let it drop? This is only making it worse."

"Oh, I do wish you'd reconsider," said Maryanne. She brought her left hand around in front of her so George could see the pistol she held. "This belonged to my husband. He showed me how to use it before he went off to the war."

George's face went white. "What the hell!" he cried.

"Quiet, George. Come inside." All business, she stepped forward and grabbed George's shoulder, pulling him into the flat. He came along, stumbling.

"Listen," said George. "I'm sorry about what happened between us, all right? But this is ridiculous."

"Get in the chair." She motioned the gun toward her ether station.

"Mrs. Spenser, please," said George, eyeing the machine with fear.

Maryanne said nothing; she pulled back the hammer on the pistol.

Slowly, George put down the briefcase and climbed into the chair.

Maryanne opened a brown bottle and poured the equivalent of four doses of the EAM cocktail into a tumbler. "Drink," she said.

"But—"

"Drink it!"

George raised the tumbler and drank down its contents.

Maryanne fastened the restraining straps over George's arms and legs and switched on the machine, turning up the gain as high as it went.

"You want to know the truth, George? I broke it off with you because I didn't like the way you pawed at me. Like you were a dog, and I was your bitch. Do you think that's what women want? Your stupid, clumsy hands all over them? Just you wait, George."

George gasped and his eyes fluttered. His breath caught in his throat and he choked. A violent tremor ran through his body.

Maryanne stepped back, unbuttoning her blouse. She felt empty. She unzipped her skirt and stepped out of it.

George lay on the ether station, his arms and legs shaking. The machine hummed. Maryanne removed the clasps from her garters and rolled down her stockings.

Finally, as the last light from the cold November day disappeared from the windows, George's body stopped moving. Maryanne, naked, gingerly removed the restraints and stared at his face. It was different, altered. His eyes opened, and Maryanne could tell at once that the eyes were not those of George Carmichael. She had succeeded.

The dark stranger rose from the chair and took her roughly in his arms. She let him kiss her mouth, her neck, her breasts. He pushed her down onto the sofa and made love to her, fantasy made flesh. She wept, her fingers digging deep scratches into his back.

Maryanne woke up in the middle of the night, alone in bed. The dark stranger was gone, the pillow cold. Her husband's pistol was missing from the kitchen table. She stood and stared at the empty table for a long while.

It was almost dawn when he returned. Maryanne was still standing in the kitchen, leaning against the stove.

"Where were you?" she said.

"Removing an obstacle," he said, placing the gun on the table. He took her hands and drew her close.

A horrid memory reared up in the bowels of Maryanne's mind, and she shrieked, pulling away.

"What's wrong?" said the dark stranger. "I am here."

Maryanne stumbled backward and leaned against the sofa. "I—I'm sorry. I thought . . . I was just remembering."

The dark stranger closed his eyes and began to unbutton George Carmichael's shirt. "Yes, the boy Parker."

"You know about him?" Maryanne's body began to shake, and she could not control it.

"I thought you knew."

"Knew what?"

"I wanted to apologize, but I can't. I'm not sorry." He slipped the shirt off, revealing a muscular, hairless chest.

"Apologize?" Maryanne sank into the sofa, clutching her arms across her breast.

"When you called to me then, I did not know how to answer your call. I was unformed then. I could not fill him."

"It was you?"

"It was."

"What do you mean? How did I call you?"

"I am yours," said the dark stranger. "I exist only to love you. You wished me into existence even before you formed the cornerstone of the palace. You sculpted me with your heart. I am for you." He fell to his knees. "Do you still not understand? In there, in your shadow palace, you dare to ask for what you cannot have in this world. You dare to love; you dare to hate. You do these things because you believe that none of it is real. You let your genuine desires take root, in the most fertile ground of the mind. I am the fruit of that desire. I am your lust, your hatred, your desire for vengeance and wickedness; and I am also your one true love, with a face and body sculpted by your own thoughts. And now I am free to clear a path for you through the waking world."

"I'm frightened." Maryanne began to cry.

"You do not need to be frightened anymore, Maryanne," he said. He sat beside her and put his arms around her. "Never again."

Maryanne wanted to scream. She wanted to pull away from him, to run out of the apartment and keep running. Instead she returned his embrace, felt the rush of him in her blood, and sighed.

Mike Resnick is the author of more than forty science fiction novels, twelve collections, over 130 stories, and two screenplays, as well as the editor of close to thirty anthologies. He has won four Hugos, and his work has been translated into twenty-two languages.

Kay Kenyon is the author of six science fiction novels. Among other themes, she has written about the transformation of Earth by an icelike ecology of information (*Maximum Ice*, nominated for the Philip K. Dick Award); the collapse of terraforming (*Rift*); and a galactic search for biodiversity (*The Seeds of Time*).

DOBCHEK, LOST IN THE FUNHOUSE

Mike Resnick and Kay Kenyon

In the morning, Dobchek found another cat in his garden. He reached under the lemon tree to grab it, but the tabby backed farther in, forcing him to scramble over the woolly thyme that had been genetically altered for winter blooms. Dobchek lunged. The cat bit. Undaunted, he latched on to the cat and hauled it out of the thyme, cursing it for digging holes in the urban garden he'd cultivated for forty years.

Holding the scrawny creature by the scruff of the neck, Dobchek strode down the apartment hallway and banged on Mrs. Murchie's door. When the door opened, he thrust the tabby at her.

"It bit me," Dobchek announced.

"You chased it, then," old Mrs. Murchie said, folding the animal into her arms. "There now," she crooned at it. Her apartment stank. Four cats and no windows will do that to a place.

"Keep them out of my garden," he insisted. "They use it for a litter box."

Dobchek had no idea how they got in. The courtyard was in the middle of the four apartment towers that formed their complex. The only way in was through his apartment. Originally the courtyard had been a gathering place for the tenants, but it had been fifty years since people dared to gather in parks. Rent on the courtyard was cheap.

Mrs. Murchie shut the door in his face. Then she opened it again and thrust her pudgy chin out. "You got a birthday, don't you Dobchek?"

He blinked, wondering how she knew. She must have memorized the birthdays of everyone in the building. It was the curse of the very old, to know too much. Especially if, like Mrs. Murchie, you weren't a scholar. She wasted her somatic computer. All that DNA computational power, squandered on the Mensa version of the *Times* crossword puzzle and raising cats.

She persisted: "Your eightieth, isn't it?"

"So?"

Eighty was the year many old people went over the top and got lost in the labyrinth of Knowing. Some people called it "lost in the funhouse." It was the opposite of Alzheimer's; you simply knew too much, and turned inward. Thinking, thinking, always thinking.

"So why not have a party?" she sneered. "How long's it been since you ever had a party, Dobchek?"

He turned and walked away from her.

But her voice followed him: "Invite all those cat-hating friends of yours." He heard the door slam.

But of course he didn't have any cat-hating friends. In point of fact, he didn't have any friends at all. But then, neither did Mrs. Murchie. He took some small measure of comfort from that.

A group of twenty-year-olds sat in their chairs, looking blankly at Dobchek. Enrolled in the engineering school, they were the brightest of the bright. (Which, because they were twenty-year-olds, was not really very bright at all.) They hadn't had time to master their somatic computers. Sure, they'd done the Change, done the treatments. But just because they now had a billion times the computing power of their grandfather's Power Macs didn't mean they understood anything.

Dobchek sighed. *Here we go again.* He tried to explain the elastic properties of ceramics using four-dimensional math, but they looked as glazed over as Mrs. Murchie's cat. There was one young woman who generally picked up on things first. She raised a hand. Thank God for her.

"Yes, Britney?"

"Will this be on the test?"

He scowled at her. A wave of annoyance rolled over him. He felt it stacking up in his throat.

Then it spilled out: "It's all on the test. You have to learn it all." They glared at him, resentful that they were young and stupid and he wasn't. "I can't spoon-feed you. You have to read. Study. Not sit there and wait for enlightenment to visit you. There is no enlightenment. Each of you is a miraculous computer. The point is to learn to access it. Not," he said, glancing at Britney, "to fool the tests."

He dismissed them, wishing he hadn't lost his temper. Their images faded like soap bubbles from bathwater, leaving him in his garden, staring at the lemon tree.

Despite the wonders of the body's somatics, the interface was damnably difficult. Subtle. Over decades, new neural connections had to be forged, as the brain learned to use the new tools at its disposal. Those new tools were the body and all its systems, guided during the Change by designer molecules, and sustained by messenger chemicals, creating links between DNA molecules or the junk DNA that was available for the task of computing. But it went beyond DNA. The biocomputer stored knowledge in organs, systems, and the gestalt of the whole body. Even the molecular programmers didn't understand how the body took to the Change and what it was becoming. It wasn't FDA-tested and -approved. It began as a guerilla movement; first a few scientists, then the very rich, followed by the kooks and the scholars, and then—well, everybody. You couldn't stop it.

But there was no magic pill of knowledge. You still had to learn the old way. Read. Study. You never forgot anything you read or heard. Just the same, kids thought they should be smarter, that four-dimensional math should be easier. Well, someday it would be. When they were old farts.

As old as Dobchek.

Feeling depressed, Dobchek took his coat from its peg and went outside. Usually he looked forward to his teleconference classes. At least, he told himself, it was some form of social contact.

He found himself walking toward the museum, his usual destination. It was twelve blocks, but he let others have the Personal Rapid Transport cars. In the time Before, he used to take the subway. Or the bus. But, being terrorist targets, those were long gone.

Long gone. Sonorous words that carried so much. He never walked up to the museum without thinking of *what* was long gone. The games at the stadium, real crowds, none of this tele-this, tele-that. Also gone—the Web, the old Internet where you could reach out and connect to friends and strangers all over the world. What a flash in the pan that was, as the terrorists figured out how to infect electronics, down to the last encrypted military computer. He remembered the day he finally threw away his computer. It was a dinosaur. Even disks carried pernicious viruses. Electronic circuits became the inroads the terrorists used to create mayhem. Silicon had let us down.

But damn, it had been fun while it lasted! He remembered meeting Alicia in the chat room, moving to e-mail love, and thence to their first real date. All that was decades ago. Alicia was a poorly remembered ghost, killed by a fundamentalist who blamed a nameless crowd at the grocery store for his country's cultural erosion via the World Wide Web.

Now Dobchek carried his computer in his creaky old body. Instead of zeros and ones, there was the A, C, T, and G of the DNA platform. Having gone beyond silicon, computers were *in vivo*. In life-forms. But there was no Net, no Web. Sure, they had keyboards and screens and silicon peripherals like printers, with cyborg connections to these. But there were no connections to other people. Not one.

All of which drove the terrorists into a frenzy. With no connections, there was simply no way they could infect the *in vivo* stuff. Two billion years of evolution had devised the best antivirus program possible: the human immune system.

Dobchek climbed the palatial steps to the museum. Flags snapped in the breeze off the river, and several people sprinted up the stairs to avoid forming a crowd at such a public place.

Dobchek took his time, letting people swarm by. A man selling pretzels at a sidewalk stand nodded at him, in friendly greeting. After a certain age, terrorists simply couldn't terrorize you anymore. You weren't anxious to die, but you were ready to. Here the guy was, selling pretzels under a big, striped umbrella. A nice target, and he couldn't care less.

In the grand vestibule, Dobchek paid his fee. A docent lectured to a group of noisy youngsters in his favorite wing, of impressionist painters. He put that off for later and turned left into the wing of those awful moving sculptures. Why couldn't art stay put anymore?

A mobile statue beckoned to him. Motion-activated, the art grouping interacted with the viewers, changing postures, changing colors. Awful stuff, really. What was art anymore, if the artists couldn't make up their minds? Dobchek didn't want to interact with art. He wanted to look at it, study it, admire it.

Next to him a woman and young boy watched a statue that was morphing into a bull. "Hey, toro!" the boy chirped. The bull's eyes glinted, perhaps getting ready to fake a charge.

The next part happened very fast, and very slow. The man from the pretzel stand was there amid the statues. Carefully unzipping his jacket, he revealed a machine attached to his chest. As he pushed a button, his body came apart. Not cleanly, like the evolving statues, but with the messy red pulp of the truly *in vivo*.

The blast knocked Dobchek and the woman into the wall. People and art lay in pieces. Shaking, the woman clung to Dobchek, and he to her. Between them was the little boy, silent, rolled into a tiny ball. Nearby a woman moaned, "Oh no, oh no!" looking at her shredded hand.

Alarms mixed with shouts and screams as museum staff rushed to help the injured. Statues began moving again, covered in blood, dancing now, a dance of death. A trickle of blood wound past them like a river seeking the sea. Sirens began whining.

Dobchek still held the woman. She moaned, looking around at the carnage. The boy said, his voice like the bark of a tiny bird, "Shussh, Mom, it's over now." She shivered, clutching Dobchek. He wrapped his fingers around her hand, and around the boy's. They were intertwined like strands of ivy, and shaking in unison.

It was in that moment that Dobchek got lost in the funhouse.

As he thought of it later, it was like falling through a hole—a hole in his self-regard. He went to a place of warm intensity. It was too fine to call peace, although later he thought it might be relief. He was so close to the woman's cheek. Her skin was pink and fragile, her eyes lustrous. And in those moments he thought that he knew her, and her son. And, strangely, that he knew himself through them. All this was impossible, delusional. But he held them tighter, squeezing out the essence of the world he'd just fallen to.

"Are you folks all right?" The medical squad guy was crouching next to them.

Dobchek blinked. The fellow was pasty-faced and cold, more like an art statue than a living being. By comparison.

The woman scrambled to her feet, shaking off Dobchek as though he were the terrorist. He held out a hand to her, wanting to prolong the moment, to say, *What happened? How do I know you?* But, narrowing her eyes, she recoiled from his hand, from his obvious neediness. She dragged the boy away, fleeing the museum.

He staggered after her, then stopped himself. Someone gave him a cloth to wipe blood off his face. He'd forgotten the terrorist blast. He was still staggered by the hole, the place where he'd fallen and dwelled for a few glorious seconds. It wasn't the funhouse, that place of computational madness. It was someplace else.

Standing at the top of the great staircase, he looked out at the mundane world. The striped umbrella of the pretzel stand now stood empty of its vendor. The woman and the boy had disappeared into the urban maze.

He missed them.

Dobchek fretted. Within three days he was back at the museum, lying in wait for the woman. The blood, debris, and pretzel stand were gone. Everything looked normal, but normal had become a deep question for Dobchek.

The next day she and the boy showed up. He followed her, taking a Personal Rapid Transit car. He had no clear plan except to plead with her to listen to him. Because without her, how could he tell whether he was going mad?

And was he? If she said "No, I didn't feel anything," then it would confirm that he'd gone Funny. Friday was his birthday. Today was Tuesday. Nothing magical about the eightieth birthday. Supermentation didn't happen like clockwork on the eightieth. But now that he thought about it, he could see the signs: he was becoming rigid and doddering. The garden, for one thing. Crossbreeding wooly thyme, and hunting Mrs. Murchie's cats. Puttering with exotic shade groundcovers and winter-hardy lemon trees. Why not just buy lemons? Put down Astro Turf? He was slipping into knowing too much about too little. It was always the first sign of those who were about to join the legion of the Lost.

The autoguided car zipped out of the city, and he overrode its questions about Destination. He didn't know. He made his choice of exits from the guideway, depending on the woman in front of him. She was the key, somehow.

Because what if he wasn't crazy? What if there were two choices when you got old? One was the funhouse. The other was Something More. He was the first generation of oldsters after the Change. It was just possible, wasn't it, that after fifty years of experience, there was an alternative to getting Lost?

He took the last few blocks on foot. Dobchek watched as the woman and her son disappeared into a small house in an exurban development. A sad little bungalow with a garden in need of pruning.

On her front porch, he forced himself to knock before he lost his nerve, then knocked again.

She opened the door, recognized him, and started to close it in his face.

"Please," he said. "I've come all this way."

She held up a small handheld pager. "All I have to do is push a button, and the security guard will be here."

"I have no problem with that," replied Dobchek. "I mean you no harm. I just want three minutes of your time. Then I'll go away, I promise."

She occupied the doorway stiffly, all angles and resistance. Her fortyish cheeks and jaws had a few deep lines, supporting her face like girders. He wished she would smile, but why should she? It was just that he thought he knew her. Knew her loneliness, her yearning. He realized how delusional that sounded.

Clutching her security device, she waited for him to have his say and leave.

He took a deep breath. "A few days ago in the museum I felt something with you and the youngster. We were almost killed, and you tend to bond with people in circumstances like that. But I'm an old man. I've seen tragedy, and this was different."

Her face hadn't changed. His time was ticking by.

"I felt I knew you and your son. . . . He *is* your son, isn't he?" She didn't answer. He had to get on with it. "What I felt was that I knew you. That you were *somebody*. That I was somebody—somebody who mattered. And since I'd never seen you before, I thought the feeling I had was significant. But I also thought I might be getting—" He hesitated to use the term *Funny*. "I thought I might be getting delusional in my old age."

He wasn't getting through. He looked out at her straggly garden, thinking how he'd like to tidy it up a bit. "Maybe I should quit teaching if I'm too old," he said, aware that he was rambling, unable to stop it. "My employer has a policy for me in a memory care condo, in case I ever need it. Maybe it's time." He paused. "Unless you felt something. You see?"

He hadn't meant to share personal things. Why would she care? But what else was he going to say? I want to hold your hand?

"I'm sorry," she said, softening a bit, but closing the door a little farther. "I don't know what you're talking about."

From behind her came a voice: "I do."

The boy came forward to stand by his mother's side. The resemblance between them was strong: strawberry blond, straight hair. Bright brown eyes. Except the boy's ears stuck out prominently as if they were scanning for broadcasts.

Around a wad of chewing gum he said, "It was like we needed each other."

The woman frowned. "I don't think we need strangers, Russell."

She started to close the door.

"Wait!" Dobchek held out a hand to stop her, but withdrew the gesture quickly. "What if we do?"

Through a foot-wide opening, the mother watched him guardedly.

He had to give voice to his most tenuous theories. Give body, perhaps, to his fantasies. "What if," he said, "our somatics are incomplete somehow? What if"—here he was just giving form to the thought that had been nudging at him for days—"what if the smarter our bodies get, the more they long for something more?" He instantly realized that she would misinterpret that. "I don't mean in the conjugal sense," he added quickly. "I never think about that anymore." He dared a smile. "Or not very much, anyway. But what if something greater arises as the result of somatic knowing. The body—the knowledgeable body—might yearn for something like . . . like a higher communion. Even just standing together in a museum."

"No," the boy said. "You have to touch." He looked up at his mother. "Without mittens."

The woman's face was closing down. He saw that she was upset by his standing on her porch, by the boy's intrusion, perhaps by Dobchek's pitiful story. But when he looked more closely, he saw tears gathering along the lids of her eyes. She gazed at him, not bothering to wipe them.

"Yes," she said. "Yes, I felt it."

He thought that this was the answer to the question he'd first asked her. He took it for his answer, the one that meant he wasn't Funny or Lost after all.

But then, what was he?

"My name's Tara," she said, extending her hand.

It was Friday, March 30, 2061. Dobchek's eightieth birthday.

And he was having a party.

Dobchek fussed with the plate of cheese and crackers. Put it in the middle of the table. Too isolated. Put it on the corner. Too self-conscious. Put it on the counter. Yes, casual elegance.

Overall, the apartment looked tacky. Only the garden made it decent. He risked opening the door to the courtyard. The afternoon sun, surprisingly warm, peeked over the apartment roof, sending a brief shaft of sunlight into the woolly thyme.

He saw a movement in the bushes. The cats. He sighed, imagining scratched-up piles of his genetically designed thyme. Russell charged out the door. "Hey, Mr. Dobchek—cats!"

Tara smiled. "He loves animals."

"Those aren't animals. Those are cats." Dobchek glanced at the clock. It was a quarter after. No one else was coming.

He should have invited more people, there were always no-shows for parties, as far as he remembered from the time Before. But the guest list had been a strain anyway: the Chief of Curriculum at the college, his dentist, the mail carrier. And at the last moment, Mrs. Murchie, because it had been her idea to have a party, even if it was a joke.

Tara settled herself at the kitchen table, looking out the door into the garden. "It's all so tidy. Very nice."

Dobchek saw it through her eyes: overshaped, overtended. A square of sunlight fell on the flagstones, creating a patch of color. That was a nice surprise. Perhaps surprise was part of what was missing from his life. From all of their lives.

He sat down at the dinette set. "I've been thinking," he began.

A silly beginning. Old people were always thinking. But he stumbled on, feeling Tara's permission to speak, to fumble into what he knew.

"I'm the first, you know. I'm in the vanguard, the first wave of people to hit old, old age. I was thirty when I did the Change. But I grew from a different soil than you did. I can remember the days when the world had a Net. When it was all connected. Not perfectly, but instantaneously, chaotically." He smiled nostalgically. "It was quite something."

Russell was crawling through the periwinkle, calling "Here, kitty kitty!"

He went on: "Maybe my mind still harbors some predilection for those days. Maybe it's filling the gap."

"Gap?" Tara asked, tracing a tea stain on the dinette set with her hand.

"It wouldn't be like the old Net. Just a limited connection, but maybe a more profound one. Maybe it's just filling the niche that exists. It isn't solving a problem. Evolution isn't directed toward anything, it just exploits what's at hand."

"Evolution?" The lines were back in her face.

"Not in the classic sense. But what if people my age are starting to enter a metamorphosis that makes them able to connect with others chemically, through touch?"

He glanced at her hands resting on the table.

She saw him looking. Smiled. "So you think it's just your generation?"

"I don't know. Maybe our memories of the old Net, of not being so damned isolated and alone, make it more likely that the elderly will make the leap."

"But we *all* felt something," she said. "Even Russell."

Dobchek nodded. "I've been thinking about that. What if the old are the catalysts? What if we can initiate—perhaps lead? And suppose it's a change everyone will eventually go through, at least if they're lonely enough."

"Like me?" she asked.

Dobchek watched Russell chasing through the undergrowth. "Or open enough to surprise."

A knock at the door.

Party guests? Dobchek rose, hurrying to open the door. There stood Mrs. Murchie. She shouldered past him. "I brought some of the neighbors," she announced.

And sure enough, Dobchek recognized the two others: Mr. Kaku and Mrs. Lessinger. He nodded at them. He'd nodded at them for forty years in the hallway, and at the mailbox, and never knew them.

They stood in his kitchen, not touching the food. Most people preferred store-wrapped food, which made sense: the terrorists liked to use poison.

Looking at the room full of people, Dobchek felt strangely emotional. No one had stood in his kitchen since Alicia died. Too long, too long. Perhaps all this talk about something more was nothing other than the yearning for simple human company. Yet he thought it went beyond that.

"Got one!" Russell shouted, cradling a cat in his arms.

The guests filed outside to see the cat, to inspect the garden. The winter afternoon was shirtsleeve warm. Everyone seemed happy to discover the garden, maybe happy to be at a real party instead of a teleconference.

Mrs. Murchie stepped off the flagstone path and traipsed over the thyme. "Springy," she said, a smile denting the side of her face. It gave Dobchek a stab of regret, but he let it go. The thyme would toughen up with use.

The group watched as afternoon shade fell down the long shaft of the bricked-in courtyard. They were standing roughly in a circle. Mrs. Murchie said, "We didn't bring any gifts," acknowledging in her way that it was a birthday party.

Russell was standing in the kitchen doorway, munching cheese. "I thought we were going to hold hands," he blurted.

Everyone looked at Dobchek, as though he should shush the boy.

Instead, Dobchek waved Russell over to his side.

"If you will indulge an old man," he said, "I'd like to give Russell's suggestion a run." He took a deep breath, and decided to risk looking stupid. Or Funny.

"Sometimes," he continued, "an old man gets a little lonely. So, if you wouldn't mind too much, I'd just like us all to hold hands for a moment."

The six of them seemed frozen like statues in a long-forgotten garden. He knew his idea was unexpected, and not exactly welcome.

He felt happy, though. Stimulated and uncertain. Alive. As though they were setting out on a sea crossed by winds from far latitudes. As the oldest one here, he'd be the navigator. Or the bird, seen far out to sea, leading them to land. But what land?

He expected to be surprised.

Tara smiled at the group, and shrugged in a disarming manner. She held out her hands to Mr. Kaku and Mrs. Murchie. Slowly, Mrs. Murchie clasped that hand. And then Dobchek's hand. Putting down the cat, Russell came forward to join in.

Then one by one, they took each other's hands. It was a fragile ring, of nobody very special, just a couple of oldsters joined by the curious and the lonely, drawn to a garden amid apartment towers. But they reached around the circle, closing the gaps between them. And as they did so, Dobchek felt it once again, that long-gone sense of connection, tentative and yet vital, part cognition, part emotion. As though this circle were somehow a community.

And perhaps now it was.

Charles Stross is a Hugo- and Nebula-award shortlisted SF writer and tech jounalist. He is the author of the short story collection *Toast* (Cosmos Books, 2002) and the novel *Singularity Sky* (Ace, August 2003). His forthcoming novels are *The Atrocity Archive* (Golden Gryphon, February 2004), *The Iron Sunrise* (Ace, August 2004), *A Family Trade* (Tor, December 2004) and *The Clan Corporate* (Tor, December 2005). He lives in Edinburgh, Scotland.

ROGUE FARM

Charles Stross

It was a bright, cool March morning: mare's tails trailed across the southeastern sky toward the rising sun. Joe shivered slightly in the driver's seat as he twisted the starter handle on the old front loader he used to muck out the barn. Like its owner, the ancient Massey Ferguson had seen better days; but it had survived worse abuse than Joe routinely handed out. The diesel clattered, spat out a gobbet of thick blue smoke, and chattered to itself dyspeptically. His mind as blank as the sky above, Joe slid the tractor into gear, raised the front scoop, and began turning it toward the open doors of the barn—just in time to see an itinerant farm coming down the road.

"Bugger," swore Joe. The tractor engine made a hideous grinding noise and died. He took a second glance, eyes wide, then climbed down from the tractor and trotted over to the kitchen door at the side of the farmhouse. "Maddie!" he called, forgetting the two-way radio clipped to his sweater hem. "Maddie! There's a farm coming!"

"Joe? Is that you? Where are you?" Her voice wafted vaguely from the bowels of the house.

"Where are you?" he yelled back.

"I'm in the bathroom."

"Bugger," he said again. "If it's the one we had round the end last month . . ."

The sound of a toilet sluiced through his worry. It was followed by a drumming of feet on the staircase; then Maddie erupted into the kitchen. "Where is it?" she demanded.

"Out front, about a quarter mile up the lane."

"Right." Hair wild and eyes angry about having her morning ablutions cut short, Maddie yanked a heavy green coat on over her shirt. "Opened the cupboard yet?"

"I was thinking you'd want to talk to it first."

"Too right I want to talk to it. If it's that one that's been lurking in the copse near Edgar's pond, I got some *issues* to discuss with it." Joe shook his head at her anger and went to unlock the cupboard in the back room. "You take the shotgun and keep it off our property," she called after him. "I'll be out in a minute."

Joe nodded to himself, then carefully picked out the twelve-gauge and a preloaded magazine. The gun's power-on self-test lights flickered erratically, but it seemed to have a full charge. Slinging it, he locked the cupboard carefully and went back out into the farmyard to warn off their unwelcome visitor.

The farm squatted, buzzing and clicking to itself, in the road outside Armitage End. Joe eyed it warily from behind the wooden gate, shotgun under his arm. It was a medium-size one, probably with half a dozen human components subsumed into it—a formidable collective. Already it was deep into farm-fugue, no longer relating very clearly to people outside its own communion of mind. Beneath its leathery black skin he could see hints of internal structure, cytocellular macroassemblies flexing and glooping in disturbing motions. Even though it was only a young adolescent, it was already the size of an antique heavy tank, and it blocked the road just as efficiently as an Apatosaurus would have. It smelled of yeast and gasoline.

Joe had an uneasy feeling that it was watching him. "Buggerit, I don't have time for this," he muttered. The stable waiting for the small herd of cloned spidercows cluttering up the north paddock was still knee-deep in manure, and the tractor seat wasn't getting any warmer while he shivered out here, waiting for Maddie to come and sort this thing out. It wasn't a big herd, but it was as big as his land and his labor could manage—the big biofabricator in the shed could assemble mammalian livestock faster than he could feed them up and sell them with an honest HAND-RAISED NOT VAT-GROWN label. "What do you want with us?" he yelled up at the gently buzzing farm.

"Brains, fresh brains for Baby Jesus," crooned the farm in a warm contralto, startling Joe half out of his skin. "Buy my brains!" Half a dozen disturbing cauliflower shapes poked suggestively out of the farm's back and then retracted again, coyly.

"Don't want no brains around here," Joe said stubbornly, his fingers whitening on the stock of the shotgun. "Don't want your kind round here, neither. Go away."

"I'm a nine-legged semiautomatic groove machine!" crooned the farm. "I'm on my way to Jupiter on a mission for love! Won't you buy my brains?" Three curious eyes on stalks extruded from its upper glacis.

"Uh—" Joe was saved from having to dream up any more ways of saying "fuck off" by Maddie's arrival. She'd managed to sneak her old battle dress home after a stint keeping the peace in Mesopotamia twenty ago, and she'd managed to keep herself in shape enough to squeeze inside. Its left knee squealed ominously when she walked it about, which wasn't often, but it still worked well enough to manage its main task—intimidating trespassers.

"You." She raised one translucent arm, pointed at the farm. "Get off my land. *Now*."

Taking his cue, Joe raised his shotgun and thumbed the selector to full auto. It wasn't a patch on the hardware riding Maddie's shoulders, but it underlined the point.

The farm hooted: "Why don't you love me?" it asked plaintively.

"Get orf my land," Maddie amplified, volume cranked up so high that Joe winced. *"Ten seconds! Nine! Eight—"* Thin rings sprang out from the sides of her arms, whining with the stress of long disuse as the Gauss gun powered up.

"I'm going! I'm going!" The farm lifted itself slightly, shuffling backwards. "Don't understand. I only wanted to set you free to explore the universe. Nobody wants to buy my fresh fruit and brains. What's wrong with the world?"

They waited until the farm had retreated round the bend at the top of the hill. Maddie was the first to relax, the rings retracting back into the arms of her battle dress, which solidified from ethereal translucency to neutral olive drab as it powered down. Joe safed his shotgun. "Bastard," he said.

"Fucking-A." Maddie looked haggard. "That was a bold one." Her face was white and pinched-looking, Joe noted. Her fists were clenched. She had the shakes, he realized without surprise. Tonight was going to be another major nightmare night, and no mistake.

"The fence." On again and off again for the past year they'd discussed wiring up an outer wire to the CHP baseload from their little methane plant.

"Maybe this time. Maybe." Maddie wasn't keen on the idea of frying passers-by without warning, but if anything might bring her around, it would be the prospect of being overrun by a bunch of rogue farms. "Help me out of this, and I'll cook breakfast," she said.

"Got to muck out the barn," Joe protested.

"It can wait on breakfast," Maddie said shakily. "I need you."

"Okay." Joe nodded. She was looking bad; it had been a few years since her last fatal breakdown, but when Maddie said "I need you," it was a bad idea to ignore her. That way led to backbreaking labor on the biofab and loading her backup tapes into the new body; always a messy business. He took her arm and steered her toward the back porch. They were nearly there when he paused.

"What is it?" asked Maddie.

"Haven't seen Bob for a while," he said slowly. "Sent him to let the cows into the north paddock after milking. Do you think—?"

"We can check from the control room," she said tiredly. "Are you really worried? . . ."

"With that thing blundering around? What do *you* think?"

"He's a good working dog," Maddie said uncertainly. "It won't hurt him. He'll be all right; just you page him."

After Joe helped her out of her battle dress, and after Maddie spent a good long while calming down, they breakfasted on eggs from their own hens, homemade cheese, and toasted bread made with rye from the hippie commune on the other side of the valley. The stone-floored kitchen in the dilapidated house they'd squatted and rebuilt together over the past twenty years was warm and homely. The only purchase from outside the valley was the coffee, beans from a hardy GM strain that grew like a straggling teenager's beard all along the Cumbrian hilltops. They didn't say much: Joe, because he never did, and Maddie, because there wasn't anything that she wanted to discuss. Silence kept her personal demons down. They'd known each other for many years, and even when there wasn't anything to discuss, they could cope with each other's silence. The voice radio on the windowsill opposite the cast-iron stove stayed off, along with the TV set hanging on the wall next to the fridge. Breakfast was a quiet time of day.

"Dog's not answering," Joe commented over the dregs of his coffee.

"He's a good dog." Maddie glanced at the yard gate uncertainly. "You afraid he's going to run away to Jupiter?"

"He was with me in the shed." Joe picked up his plate and carried it to the sink, began running hot water onto the dishes. "After I cleaned the lines I told him to go take the herd up the paddock while I did the barn." He glanced up, looking out the window with a worried expression. The Massey Ferguson was parked right in front of the open barn doors as if holding at bay the mountain of dung, straw, and silage that mounded up inside like an invading odorous enemy, relic of a frosty winter past.

Maddie shoved him aside gently and picked up one of the walkie-talkies from the charge point on the windowsill. It bleeped and chuckled at her. "Bob, come in. Over." She frowned. "He's probably lost his headset again."

Joe racked the wet plates to dry. "I'll move the midden. You want to go find him?"

"I'll do that." Maddie's frown promised a talking-to in store for the dog when she caught up with him. Not that Bob would mind: words ran off him like water off a duck's back. "Cameras first." She prodded the battered TV set to life, and grainy bisected views flickered across the screen, garden, yard, Dutch barn, north paddock, east paddock, main field, copse. "Hmm."

She was still fiddling with the smallholding surveillance system when Joe clambered back into the driver's seat of the tractor and fired it up once more. This time there was no cough of black smoke, and as he hauled the mess of manure out of the barn and piled it into a three-meter-high midden, a quarter of a ton at a time, he almost managed to forget about the morning's unwelcome visitor. Almost.

By late morning, the midden was humming with flies and producing a remarkable stench, but the barn was clean enough to flush out with a hose and broom. Joe was about to begin hauling the midden over to the fermentation tanks buried round the far side of the house when he saw Maddie coming back up the path, shaking her head. He knew at once what was wrong.

"Bob," he said, expectantly.

"Bob's fine. I left him riding shotgun on the goats." Her expression was peculiar. "But that *farm*—"

"Where?" he asked, hurrying after her.

"Squatting in the woods down by the stream," she said tersely. "Just over our fence."

"It's not trespassing, then."

"It's put down feeder roots! Do you have any idea what that means?"

"I don't—" Joe's face wrinkled in puzzlement. "Oh."

"Yes. *Oh*." She stared back at the outbuildings between their home and the woods at the bottom of their smallholding, and if looks could kill, the intruder would be dead a thousand times over. "It's going to estivate, Joe, then it's going to grow to maturity on our patch. And do you know where it said it was going to go when it finishes growing? Jupiter!"

"Bugger," Joe said faintly, as the true gravity of their situation began to sink in. "We'll have to deal with it first."

"That wasn't what I meant," Maddie finished. But Joe was already on his way out the door. She watched him crossing the yard, then shook her head. "Why am I stuck here?" she asked, but the cooker wasn't answering.

The hamlet of Outer Cheswick lay four kilometers down the road from Armitage End, four kilometers past mostly derelict houses and broken-down barns, fields given over to weeds and walls damaged by trees. The first half of the twenty-first century had been cruel years for the British agrobusiness sector; even harsher if taken in combination with the decline in population and the consequent housing surplus. As a result, the dropouts of the forties and fifties were able to take their pick from among the gutted shells of once fine farmhouses. They chose the best and moved in, squatted in the derelict outbuildings, planted their seeds and tended their flocks and practiced their DIY skills, until a generation later a mansion fit for a squire stood in lonely isolation alongside a decaying road where no more cars drove. Or rather, it would have taken a generation had there been any children against whose lives it could be measured; these were the latter decades of the population crash, and what a previous century would have labeled downshifter DINK couples were now in the majority, far outnumbering any breeder colonies. In this aspect of their life, Joe and Maddie were boringly conventional. In other respects they weren't: Maddie's nightmares, her aversion to alcohol, and her withdrawal from society were all relics of her time in Peaceforce. As for Joe, he liked it here. Hated cities, hated the Net, hated the burn of the new. Anything for a quiet life . . .

The Pig and Pizzle, on the outskirts of Outer Cheswick, was the only pub within about ten kilometers—certainly the only one within staggering distance for Joe when he'd had a skinful of mild—and it was naturally a seething den of local gossip, not least because Ole Brenda refused to allow electricity, much less bandwidth, into the premises. (This was not out of any sense of misplaced technophobia, but a side effect of Brenda's previous life as an attack hacker with the European Defense Forces.)

Joe paused at the bar. "Pint of bitter?" he asked tentatively. Brenda glanced at him and nodded, then went back to loading the antique washing machine. Presently she pulled a clean glass down from the shelf and held it under the tap.

"Hear you've got farm trouble," she said noncommitally as she worked the hand pump on the beer engine.

"Uh-huh." Joe focused on the glass. "Where'd you hear that?"

"Never you mind." She put the glass down to give the head time to settle. "You want to talk to Arthur and Wendy-the-Rat about farms. They had one the other year."

"Happens." Joe took his pint. "Thanks, Brenda. The usual?"

"Yeah." She turned back to the washer. Joe headed over to the far corner where a pair of huge leather sofas, their arms and backs ripped and scarred by generations of Brenda's semiferal cats, sat facing each other on either side of a cold hearth. "Art, Rats. What's up?"

"Fine, thanks." Wendy-the-Rat was well over seventy, one of those older folks who had taken the p53 chromosome hack and seemed to wither into timelessness: white dreadlocks, nose and ear studs dangling loosely from leathery holes, skin like a desert wind. Art had been her boy-toy once, back before middle age set its teeth into him. He hadn't had the hack, and looked older than she did. Together they ran a smallholding, mostly pharming vaccine chicks but also doing a brisk trade in high-nitrate fertilizer that came in on the nod and went out in sacks by moonlight.

"Heard you had a spot of bother?"

" 'S true." Joe took a cautious mouthful. "Mm, good. You ever had farm trouble?"

"Maybe." Wendy looked at him askance, slitty-eyed. "What kinda trouble you got in mind?"

"Got a farm collective. Says it's going to Jupiter or something. Bastard's homesteading the woods down by Old Jack's stream. Listen . . . Jupiter?"

"Aye, well, that's one of the destinations, sure enough." Art nodded wisely, as if he knew anything.

"Naah, that's bad." Wendy-the-Rat frowned. "Is it growing trees, do you know?"

"Trees?" Joe shook his head. "Haven't gone and looked, tell the truth. What the fuck makes people do that to themselves, anyway?"

"Who the fuck cares?" Wendy's face split in a broad grin. "Such as don't think they're human anymore, meself."

"It tried to sweet-talk us," Joe said.

"Aye, they do that," said Arthur, nodding emphatically. "Read somewhere they're the ones as think we aren't fully human. Tools an' clothes and farmyard machines, like? Sustaining a pre-post-industrial lifestyle instead of updating our genome and living off the land like God intended?"

" 'Ow the hell can something with nine legs and eye stalks call itself human?" Joe demanded, chugging back half his pint in one angry swallow.

"It used to be, once. Maybe used to be a bunch of people." Wendy got a weird and witchy look in her eye. " 'Ad a boyfriend back thirty, forty years ago, joined a Lamarckian clade. Swapping genes an' all, the way you or me'd swap us underwear. Used to be a 'viromentalist back when antiglobalization was about big corporations pissing on us all for profits. Got into gene hackery and self-sufficiency big time. I slung his fucking ass when he turned green and started photosynthesizing."

"Bastards," Joe muttered. It was deep green folk like that who'd killed off the agricultural-industrial complex in the early years of the century, turning large portions of the countryside into ecologically devastated wilderness gone to rack and ruin. Bad enough that they'd set millions of countryfolk out of work—but that they'd gone on to turn green, grow extra limbs and emigrate to Jupiter orbit was adding insult to injury. And having a good time in the process, by all accounts. "Din't you 'ave a farm problem, coupla years back?"

"Aye, did that," said Art. He clutched his pint mug protectively.

"It went away," Joe mused aloud.

"Yeah, well." Wendy stared at him cautiously.

"No fireworks, like." Joe caught her eye. "And no body. Huh."

"Metabolism," said Wendy, apparently coming to some kind of decision. "That's where it's at."

"Meat—" Joe, no biogeek, rolled the unfamiliar word around his mouth irritably. "I used to be a software dude before I burned, Rats. You'll have to 'splain the jargon 'fore using it."

"You ever wondered how those farms *get* to Jupiter?" Wendy probed.

"Well." Joe shook his head. "They, like, grow stage trees? Rocket logs? An' then they est-ee-vate and you are fucked if they do it next door 'cause when those trees go up they toast about a hundred hectares?"

"Very good," Wendy said heavily. She picked up her mug in both hands and gnawed on the rim, edgily glancing around as if hunting for police gnats. "Let's you and me take a hike."

Pausing at the bar for Ole Brenda to refill her mug, Wendy led Joe out past Spiffy Buerke—throwback in green wellingtons and Barbour jacket—and her latest femme, out into what had once been a car park and was now a tattered wasteground out back behind the pub. It was dark, and no residual light pollution stained the sky: the Milky Way was visible overhead, along with the pea-size red cloud of orbitals that had gradually swallowed Jupiter over the past few years. "You wired?" asked Wendy.

"No, why?"

She pulled out a fist-size box and pushed a button on the side of it, waited for a light on its side to blink green, and nodded. "Fuckin' polis bugs."

"Isn't that a—?"

"Ask me no questions, an' I'll tell you no fibs." Wendy grinned.

"Uh-huh." Joe took a deep breath: he'd guessed Wendy had some dodgy connections, and this—a portable local jammer—was proof: any police bugs within two or three meters would be blind and dumb, unable to relay their chat to the keyword-trawling subsentient coppers whose job it was to prevent conspiracy-to-commit offenses before they happened. It was a relic of the Internet Age, when enthusiastic legislators had accidentally demolished the right of free speech in public by demanding keyword monitoring of everything within range of a network terminal—not realizing that in another few decades 'network terminals' would be self-replicating 'bots the size of fleas and about as common as dirt. (The Net itself had collapsed shortly thereafter, under the weight of self-replicating viral libel lawsuits, but the legacy of public surveillance remained.) "Okay. Tell me about metal, meta—"

"Metabolism." Wendy began walking toward the field behind the pub. "And stage trees. Stage trees started out as science fiction, like? Some guy called Niven—anyway. What you do is, you take a pine tree and you hack it. The xylem vessels running up the heartwood, usually they just lignify and die, in a normal tree. Stage trees go one better, and before the cells die, they *nitrate* the cellulose in their walls. Takes one fuckin' crazy bunch of hacked 'zymes to do it, right? And lots of energy, more energy than trees'd normally have to waste. Anyways, by the time the tree's dead, it's like ninety percent nitrocellulose, plus built-in stiffeners and baffles and microstructures. It's not, like, straight explosive—it detonates cell by cell, and *some* of the xylem tubes are, eh, well, the farm grows custom-hacked fungal hyphae with a depolarizing membrane nicked from human axons down them to trigger the reaction. It's about efficient as 'at old-time Ariane or Atlas rocket. Not very, but enough."

"Uh." Joe blinked. "That meant to mean something to me?"

"Oh 'eck, Joe." Wendy shook her head. "Think I'd bend your ear if it wasn't?"

"Okay." He nodded, seriously. "What can I do?"

"Well." Wendy stopped and stared at the sky. High above them, a belt of faint light sparkled with a multitude of tiny pinpricks; a deep green wagon train making its orbital transfer window, self-sufficient posthuman Lamarckian colonists, space-adapted, embarking on the long, slow transfer to Jupiter.

"Well?" He waited expectantly.

"You're wondering where all that fertilizer's from," Wendy said elliptically.

"Fertilizer." His mind blanked for a moment.

"Nitrates."

He glanced down, saw her grinning at him. Her perfect fifth set of teeth glowed alarmingly in the greenish overspill from the light on her jammer box.

"Tha' knows it make sense," she added, then cut the jammer.

When Joe finally staggered home in the small hours, a thin plume of smoke was rising from Bob's kennel. Joe paused in front of the kitchen door and sniffed anxiously, then relaxed. Letting go of the door handle, he walked over to the kennel and sat down outside. Bob was most particular about his den—even his own humans didn't go in there without an invitation. So Joe waited.

A moment later there was an interrogative cough from inside. A dark, pointed snout came out, dribbling smoke from its nostrils like a particularly vulpine dragon. "Rrrrrrr?"

" 'S'me."

"Uuurgh." A metallic click. "Smoke good smoke joke cough tickle funny arf arf?"

"Yeah, don't mind if I do."

The snout pulled back into the kennel; a moment later it reappeared, teeth clutching a length of hose with a mouthpiece on one end. Joe accepted it graciously, wiped off the mouthpiece, leaned against the side of the kennel, and inhaled. The weed was potent and smooth: within a few seconds the uneasy dialogue in his head was still.

"Wow, tha's a good turnup."

"Arf-arf-ayup."

Joe felt himself relaxing. Maddie would be upstairs, snoring quietly in their decrepit bed: waiting for him, maybe. But sometimes a man just had to be alone with his dog and a good joint, doing man-and-dog stuff. Maddie understood this and left him his space. Still . . .

" 'At farm been buggering around the pond?"

"Growl exclaim fuck-fuck yup! Sheep-shagger."

"If it's been at our lambs—"

"Nawwwwwrr. Buggrit."

"So whassup?"

"Grrrr, Maddie yap-yap farmtalk! Sheep-shagger."

"Maddie's been *talking* to it?"

"Grrr yes-yes!"

"Oh, shit. Do you remember when she did her last backup?"

The dog coughed fragrant blue smoke. "Tank thump-thump full cow moo beef clone."

"Yeah, I think so, too. Better muck it out tomorrow. Just in case."

"Yurrrrrp." But while Joe was wondering whether this was agreement or just a canine eructation a lean paw stole out of the kennel mouth and yanked the hookah back inside. The resulting slobbering noises and clouds of aromatic blue smoke left Joe feeling a little queasy: so he went inside.

The next morning, over breakfast, Maddie was even quieter than usual. Almost meditative.

"Bob said you'd been talking to that farm," Joe commented over his eggs.

"Bob—" Maddie's expression was unreadable. "Bloody dog." She lifted the Rayburn's hot plate lid and peered at the toast browning underneath. "Talks too much."

"Did you?"

"Ayup." She turned the toast and put the lid back down on it.

"Said much?"

"It's a farm." She looked out the window. "Not a fuckin' worry in the world 'cept making its launch window for Jupiter."

"It—"

"Him. Her. They." Maddie sat down heavily in the other kitchen chair. "It's a collective. Usedta be six people. Old, young, whatever, they's decided ter go to Jupiter. One of 'em was telling me how it happened. How she'd been living like an accountant in Bradford, had a nervous breakdown. Wanted *out*. Self-sufficiency." For a moment her expression turned bleak. "Felt herself growing older but not bigger, if you follow."

"So how's turning into a bioborg an improvement?" Joe grunted, forking up the last of his scrambled eggs.

"They're still separate people: bodies are overrated, anyway. Think of the advantages: not growing older, being able to go places and survive anything, never being on your own, not bein' trapped—" Maddie sniffed. "Fuckin' toast's on fire!"

Smoke began to trickle out from under the hot plate lid. Maddie yanked the wire toasting rack out from under it and dunked it into the sink, waited for waterlogged black crumbs to float to the surface before taking it out, opening it, and loading it with fresh bread.

"Bugger," she remarked.

"You feel trapped?" Joe asked. *Again?* He wondered.

Maddie grunted evasively. "Not your fault, love. Just life."

"Life." Joe sniffed, then sneezed violently as the acrid smoke tickled his nose. "Life!"

"Horizon's closing in," she said quietly. "Need a change of horizons."

"Ayup, well, rust never sleeps, right? Got to clean out the winter stables, haven't I?" said Joe. He grinned uncertainly at her as he turned away. "Got a shipment of fertilizer coming in."

In between milking the herd, feeding the sheep, mucking out the winter stables, and surreptitiously EMPing every police 'bot on the farm into the silicon afterlife, it took Joe a couple of days to get round to running up his toy on the household fabricator. It clicked and whirred to itself like a demented knitting machine as it ran up the gadgets he'd ordered—a modified crop sprayer with double-walled tanks and hoses, an air rifle with a dart loaded with a potent cocktail of tubocurarine and etorphine, and a breathing mask with its own oxygen supply.

Maddie made herself scarce, puttering around the control room but mostly disappearing during the daytime, coming back to the house after dark to crawl, exhausted, into bed. She didn't seem to be having nightmares, which was a good sign. Joe kept his questions to himself.

It took another five days for the smallholding's power field to concentrate enough juice to begin fueling up his murder weapons. During this time, Joe took the house off-Net in the most deniable and surreptitiously plausible way, a bastard coincidence of squirrel-induced cable fade and a badly shielded alternator on the backhoe to do for the wireless chitchat. He'd half expected Maddie to complain, but she didn't say anything—just spent more time away in Outer Cheswick or Lower Gruntlingthorpe or wherever she'd taken to holing up.

Finally, the tank was filled. So Joe girded his loins, donned his armor, picked up his weapons, and went to do battle with the dragon by the pond.

The woods around the pond had once been enclosed by a wooden fence, a charming copse of old-growth deciduous trees, elm and oak and beech growing uphill, smaller shrubs nestling at their ankles in a green skirt that reached all the way to the almost-stagnant waters. A little stream fed into it during rainy months, under the feet of a weeping willow; children had played here, pretending to explore the wilderness beneath the benevolent gaze of their parental control cameras.

That had been long ago. Today the woods really *were* wild. No kids, no picnicking city folks, no cars. Badgers and wild coypu and small, frightened wallabies roamed the parching English countryside during the summer dry season. The water drew back to expose an apron of cracked mud, planted with abandoned tin cans and a supermarket trolley of Precambrian vintage, its GPS tracker long since shorted out. The bones of the technological epoch, poking from the treacherous surface of a fossil mud bath. And around the edge of the mimsy puddle, the stage trees grew.

Joe switched on his jammer and walked in among the spear-shaped conifers. Their needles were matte black and fuzzy at the edges, fractally divided, the better to soak up all the available light: a network of taproots and fuzzy black grasslike stuff covered the ground densely around them. Joe's breath wheezed noisily in his ears, and he sweated into the airtight suit as he worked, pumping a stream of colorless smoking liquid at the roots of each ballistic trunk. The liquid fizzed and evaporated on contact: it seemed to bleach the wood where it touched. Joe carefully avoided the stream: this stuff made him uneasy. As did the trees, but liquid nitrogen was about the one thing he'd been able to think of that was guaranteed to kill the trees stone dead without igniting them. After all, they had cores that were basically made of gun cotton—highly explosive, liable to go off if you subjected them to a sudden sharp impact or the friction of a chainsaw. The tree he'd hit on creaked ominously, threatening to fall sideways, and Joe stepped round it, efficiently squirting at the remaining roots. Right into the path of a distraught farm.

"My holy garden of earthly delights! My forest of the imaginative future! My delight, my trees, my trees!" Eye stalks shot out and over, blinking down at him in horror as the farm reared up on six or seven legs and pawed the air in front of him. "Destroyer of saplings! Earth mother rapist! Bunny-strangling vivisectionist!"

"Back off," said Joe, dropping his cryogenic squirter and fumbling for his air gun.

The farm came down with a ground-shaking thump in front of him and stretched eyes out to glare at him from both sides. They blinked, long black eyelashes fluttering across angry blue irises. "How *dare* you?" demanded the farm. "My treasured seedlings!"

"Shut the fuck up," Joe grunted, shouldering his gun. "Think I'd let you burn my holding when tha' rocket launched? Stay the *fuck* away," he added as a tentacle began to extend from the farm's back.

"My crop," it moaned quietly. "My exile! Six more years around the sun chained to this well of sorrowful gravity before next the window opens! No brains for Baby Jesus! Defenestrator! We could have been so happy together if you hadn't fucked up! Who set you up to this? Rat Lady?" It began to gather itself, muscles rippling under the leathery mantle atop its leg cluster.

So Joe shot it.

Tubocurarine is a muscle relaxant: it paralyzes skeletal muscles, the kind over which human nervous systems typically exert conscious control. Etorphine is an insanely strong opiate—twelve hundred times as potent as heroin. Given time, a farm, with its alien adaptive metabolism and consciously controlled proteome might engineer a defense against the etorphine—but Joe dosed his dart with enough to stun a blue whale, and he had no intention of giving the farm enough time.

It shuddered and went down on one knee as he closed in on it, a Syrette raised. "Why?" it asked plaintively in a voice that almost made him wish he hadn't pulled the trigger. "We could have gone together!"

"Together?" he asked. Already the eye stalks were drooping; the great lungs wheezed effortfully as it struggled to frame a reply.

"I was going to ask you," said the farm, and half its legs collapsed under it, with a thud like a baby earthquake. "Oh, Joe, if only—"

"Joe? *Maddie?*" he demanded, nerveless fingers dropping the tranquilizer gun.

A mouth appeared in the farm's front, slurred words at him from familiar seeming lips, words about Jupiter and promises. Appalled, Joe backed away from the farm. Passing the first dead tree, he dropped the nitrogen tank: then an impulse he couldn't articulate made him turn and run, back to the house, eyes almost blinded by sweat or tears. But he was too slow, and when he dropped to his knees next to the farm, pharmacopoeia clicking and whirring to itself in his arms, he found it was already dead.

"Bugger," said Joe, and he stood up, shaking his head. *"Bugger."* He keyed his walkie-talkie: "Bob, come in, Bob!"

"Rrrrowl?"

"Momma's had another break-down. Is the tank clean, like I asked?"

"Yap!"

"Okay. I got 'er backup tapes in t'office safe. Let's get t' tank warmed up for 'er an' then shift t' tractor down 'ere to muck out this mess."

That autumn, the weeds grew unnaturally rich and green down in the north paddock of Armitage End.

Terry McGarry is a copy editor and Irish musician from New York City, with an honors degree in English from Princeton (where she wrote her graduating thesis on dystopian science fiction) and fifteen years logged in the editorial department of *The New Yorker* magazine. Her novels *Illumination* and *The Binder's Road* are available now, with a third, *Triad,* forthcoming. More than forty of her short stories have appeared in genre publications, from *Amazing Stories* to *Realms of Fantasy,* and her SF poetry is collected in the award-winning chapbook *Imprinting*.

SWIFTWATER

Terry McGarry

Outside a house of old age and loss, John Jasper watched his client, an unmemorable young woman, disappear on foot into the gaslit night. Hovering before him, like retinal burn, was her senile grandmother's expression: the brief return of comprehension to the watery eyes, the seamed lips trembling as he gave her back her wedding day, her daughter's birth. For ten years he'd saved them for her, and now disbursed them weekly. History in escrow; awed joy the dividend.

The client had asked him to find things hidden in her grandmother's decaying mind: the location of a passbook, a safe-deposit key. "I can't retrieve what's gone," he'd told her. "I can't go in and reconstruct data." The tech term was lost on her, and he realized: *Of course. She's only what, twenty-three? She was a baby when it happened. She doesn't remember what it was like before.*

//We're standing next to a water fountain, boxes all around us, breaking down the office. Larry took a twelve-story header last week. "At least the techies can translate their skills, once things calm down," I say bit-

terly. My coworker still hasn't accepted the crisis at all, still believes it's going to go away. "Code monkeys like us . . . I don't know, buddy . . ."//

A displacement of air, a sigh of pressure brakes, returned him blinking to the present.

The nursing home was off the beaten track; the street was quiet at this hour. There wasn't so much as the squeak of a commuter pedbike. When he felt the presence of the man behind him, when he felt the circular insistence of the gun barrel through his onesuit, when he saw the dark-windowed ricksha at the curb like a cruising predator, it was like a memory: a memory of how he hadn't been wary, how he hadn't paid attention, how he'd let a client's past distract him for the seconds required to take him unawares.

"Get in the ricky, please, Mr. Jasper."

The glow of the gaslamps looked stark and unfamiliar, as if he were remembering the Victorian childhood of someone immensely old.

"Don't kill me," he said. Not a random mugging, or he'd already be bleeding on the street, remembering the blow, the hands groping his pockets. The man knew his name. "Whatever it is you think I have stored in my head, you can reason with me about it, okay?"

Donna and Joey were waiting for him at home. They were supposed to talk some more about saving Joey's memories now, to return sweet childhood to him later. It would be an invasion, Jasper had said; he didn't want to know *that* much about his son; doctors don't operate on their own kids . . . but Donna saw it as a priceless, unique gift. And Joey wanted it.

Joey had no virtual interface modules to record his experiences or link him into the schoolroom of the world. For Joey, other planets—commonplace destinations in John Jasper's youth—were a fantasy, the stars forever out of reach. Jasper would bequeath him memories as anecdotes, boring stories of the old days, the legacy of any ordinary father to his boy.

The weapon moved to his neck, steel on skin.

"Now, Mr. Jasper."

He raised his hands away from his sides and complied. The runner pulled away as soon as the guy with the gun closed the door.

He couldn't keep the memories separate from his own.

He knew whose were whose; he didn't—couldn't—deliver for one client another's college graduation, or give the second the first's wedding night. He knew which ones weren't his. He wasn't losing his mind. Yet.

But all of them were part of him: currents of memory in a river of time.

Before the Crash, he'd been an anthropologist, his talent useful in gathering primary-source material. Since the Crash, a vim-starved populace craved his mental storage facilities. ESP had become big business, as people accepted the absence of tech and turned inward to the powerful vistas of the human mind. Meatware was at a premium. But he remained freelance—joined no esper guilds, sold his soul to no company. And he wasn't telepathic. All he could do was remember, if you remembered it for him first. And then he could play it back for you.

Or tell someone else.

That was the rub.

He sat back, obedient, in the ricky, and the metal O withdrew from his neck, leaving a numb imprint. *Okay,* he thought, *this is okay, I can talk to these people, we can work out an arrangement—*

The goon next to him shifted, and before he could turn, there was a sharp jab in the middle of the O, and in a rush of anesthesia that he thought might be his death coming to meet him was a glyph of anonymous past:

//an undulating cylindrical river, we are inside *a river, this is what it looks like from the inside, silver and molten ice, light and silver shadow, the most beautiful thing I have ever imagined, didn't think I could feel anything the way other people felt but this thing is a* feeling, *I am having a real live* feeling, *and I think it must be joy this swelling of my heart the tears in my eyes the ache in my belly how does anyone bear this* feeling, *this* joy//

He woke in a white-plastered room with one door, no windows, two slatted wooden chairs. His chair swayed, joints loose. In the other chair, hands folded patiently, was a Filipina.

He waited for the threats, the explanation, the executioner. She didn't say anything. He broke the ice before he got so scared he couldn't talk. "I know you can't tell me what you think I remember, because if I don't remember it, the cat would be out of the bag, right? But I do not know why I'm here."

In a soft contralto, she said, "Malcolm Kiernan."

//In C-and-C, waiting for a report of the Kip Thorne *to come back from the monitoring station. Three more minutes till Jupiter gets the hell out of the signal's way. Beth Atherton's a spaz, but she's the best damn pilot there is. . . .//*

//Two fingers of booze left in the bottle. There must be a faster way to die. A rippling form: someone come to grovel at my famous feet, "Oh Mr. Kiernan could I have your autograph it doesn't matter that there won't be any more hyperspace missions ever it doesn't matter that the Crash destroyed your life you're still tops in my book—" But no, not a sycophant, not someone come to torture me about my drowned goddamn past. . . .//

He blinked, pressed his fingers to his temples, found himself sober again, found himself *him* again, and saw the same woman sitting before him. "You're the one who approached him," he said, "about trying to—"

//The pain of bloody knuckles, a ragged hole in ancient, crumbling drywall. God damn *it! How much goddamn sport do the goddamn gods* get *at my expense? Bring me back from the brink, give me purpose, then throw a death sentence in my face, a goddamn* disease, *after everything I've been through to be betrayed by my own goddamn* cells!*//*

He was toppling. He grabbed the bottom of the chair.

"He lied to me. He told me he had been diagnosed with Korsakoff's syndrome, that his heavy drinking after the Crash had caught up with him."

He said this as if it were the woman's fault. She smiled at him without warmth.

"Why didn't you realize it when he originally gave you the memory you just experienced?" she asked mildly.

He didn't know. He probably hadn't thought about it. Businessmen, powerful men—he didn't like them. A disease is a disease; he didn't examine it to be sure what disease it was.

"So you're trying to refit the prototype hyperdrive ship, the one Kiernan worked on before the Crash." He tried for the same conversational tone she was using, but his voice shook. "He knew he was going to die before your project was completed, and so he trained his successors to be able to finish it for him. But maybe there was something . . ." He cleared his throat to push the fear down. "Kiernan slipped me some information you need, didn't he?"

She shook her head. Her bangs didn't swing; her hair looked painted on. "He told his staff everything technical they needed to know; he wrote everything down. The problem, Mr. Jasper, is merely that you know of the project at all."

He slammed the chair, caught an echo of Kiernan's stinging fist. "I wouldn't have known I knew about it if you hadn't triggered me!"

She shrugged. "You might have been triggered by something else eventually. It's a risk we can't take."

"Mad scientists." It just came out. "I've avoided cops and job-sick neurologists and company thugs, and now I'm going to be executed because I have the microfilm with the secret formula on it."

Flippancy. Flippancy, when all he wanted to say was *Please, please don't do this, I have a little boy, my wife will never know what happened to me, please—*

She rose and went to the door, her steps a whisper on another century's cracked linoleum. "You've summed up the situation quite well, Mr. Jasper. I'm sorry about this. We don't have the resources to hold you prisoner, and your talents wouldn't be of much use to us . . . unless you have lied about your ability to read minds?"

He didn't answer. His eyes burned.

"Then we have no more to discuss. Understand that the future of this planet is at stake; we're not a bunch of crazies on some harebrained crusade. You won't be dying for nothing."

"Can't you just buy me off?" Anything to belay death for one more heartbeat. "Or hold me prisoner just for a little while? Just till your ship has lifted? Or how about—?"

"We're a small group, Mr. Jasper." No regret. Pragmatism. "We have to do things in the most expedient way."

It would take only one man with strong hands or a weapon to put an end to him, and all the memories with him.

She rapped on the door.

He remembered the most beautiful things ever seen by human eyes: sunrise on Mars like the inside of a rose in first bloom, mountains of fractal teeth, oceans of frozen mercury.

He remembered exactly what the client remembered when they linked up, but he didn't have the client's associations with the memory. The subconscious underpinnings were inaccessible. Where it all came back together was in the client's mind. Even senile minds: he played back the memories for them, and *they* made the associations. Their memories would always belong to them, even entrusted into someone else's keeping.

He screened his customers. No pedophiles or rapists, no one who'd done something he didn't want to live with for the rest of his life. But people didn't tell you everything, and he'd never met an esper who could suck bad memories away.

He wouldn't want to lose them all. Most of them were the memories of the dead. Those people lived on in him, for a brief while. He respected that. He cherished it.

But he had a bunch here that he would give anything—anything but his wife, or his son, or his life—to erase. And he didn't have any idea which ones they were.

He squinched his eyes, preparing to run through every memory he possessed. He'd never done that; they came back on their own, free association. He didn't know if he *could* do it. But if he couldn't find anything, by god he would make something up. He would not believe there was no way out. He would not panic.

Too late. He had already panicked. *All right,* he thought, *use the adrenaline*.

He remembered building the Mars habitats. He remembered Luna Alpha before the explosion. He remembered the day the *Thorne* lifted off, the first manned mission through hyperspace. Hyperdrive. They were refitting a hyperdrive ship. He dug around in his head, but there weren't any clear connections, and he didn't have a search engine. He had never categorized the memories, never attached mnemonics to them. He'd never had to; the people they were meant for triggered them when the time came.

God *damn* it. It was in there somewhere—

Hyperdrive. The biggest development since the A-bomb, but its antithesis: celebrated by a world at peace. The *Thorne* mission was a success, and in an eyeblink emissaries were on their way to Theron, Bellatrix IX, Incarnatus. Terra would take its place in the universe, among new friends.

Then one of the ships brought something back, something impossible: something that crashed the unassailable network of quantum prisms that ran the world. The wireless Net could not be killed, but it was composed of artificial intelligence, and the intelligence could go insane. Power grids failed, suborbitals nosedived, banks winked out, comms spouted gibberish, vim fried synapses worldwide. There was no plug to pull. Anachronist groups, quick to jump on the excuse, went on a Calivinist Luddite rampage, smashing whatever tech they could find. But they couldn't get it all, and the death and chaos ended only when the prisms, in a final agony of madness, shut themselves down.

For a moment, Jasper thought he knew why. He remembered—a child, a programmer, an architecture of dreams . . . he didn't know whose memory it was, couldn't make sense of it. Was that what they wanted—after all these years, an explanation for what was over and done? You couldn't reconstruct the Net. Silicon, vacuum tubes—make a computer, and it self-destructed after a period of incomprehensible functioning. The *thing* was still out there, inexplicable, inexorable.

Were they looking for a cure?

The world recovered—was still recovering, slowly—as the Anachronists had dreamed: by retreating to the late nineteenth century. Mechanical things still worked. But it was like starting from scratch. A lot of people died in the riots. A lot of people killed themselves. And those exotic, exciting planets no one else would never see—Theron, Bellatrix IX—recoiled in terror.

Recoiled and protected themselves. The Sol system was put under quarantine. Ships, beacons, mines hovered at a safe distance, to keep humanity—and whatever caused the Crash—from harming the rest of the galaxy. No one worried about it. Humanity was too busy trying to survive, day by day, in a world devoid of tech. Younger people, who'd never known the near-telepathy of vim communication, the safe and certain world the prisms had tended, accepted the quarantine with a distant shrug. They no longer cared about the stars, or the rivers of quicksilver hyperspace that could carry them there.

But these people did.

These people were going to try to run the blockade.

Without computers.

He knew it in a moment of perfect clarity, as the fragments of memory swirled and mixed in the neural matrix of his brain. What use, otherwise, in refitting a hypership? But there was no way to navigate, not without tech . . . and even if you could substitute mechanical devices for electronic, you couldn't open the gate. . . .

There had been eight missions in all, and all but one had gone smoothly.

All but one . . .

He stands in a hospice room, with floral-print wallpaper pasted over the dead wallscreens, a scent of disinfectant pasted over the yeasty odor of incontinence.

His client lies on the bed. Her body is as small as a child's under the covers. Her ruined eyes are sunken pits in a face of ethereal beauty.

Katerina Thaïs, a pilot, was rendered blind, mute, and quadriplegic by an unexplained accident on the eighth manned hyperspace mission. Before the interface was complete for her prosthetics, the Crash wiped out the software, and her family approached him with an idea for communicating with her. She would make a mental event of things she wanted them to know, then give the memory of it to him.

Although he's not an empath, he's hesitant to touch her crushing pain and isolation. But he lays his fingertips on her brow and accepts her memories. They are disjointed, alien . . . for a moment, before he's lost his self, he is awed at this otherness. . . .

//I'm looking out the forward viewscreen, but not at what the viewscreen shows. I just did that weird thing with my mind, like crossing my eyes. It worked it worked it worked! So beautiful. So beautiful. The shifting patterns. The silver and light not silver not light. I feel home on the other side, I sense the familiar stars in their correct places.

//The AI is babbling at me. It has the tone of voice that frightened people have. I think it is like me. I think it is programmed to feign fright it doesn't feel. "Shut up!" I swat at the interface. "Shut up shut up!" It distracts me with warning lights. Red and green and red and blue. They are daggers in my eyes. The warnings are icepicks in my ears. "I can do this! I am doing this! Shut up!"

//My concentration is gone. I have to look in, not out, and the machine forces me to pay attention to it.

//I scream. It is not the right time. I can't let go . . .//

He stumbles away from her, breaking contact before she's finished.

Her mother cries out; she must think it was a memory of intense pain that drove him back. But it isn't that at all.

He steps back up to her. He is too small and plain a human shell to absorb what she has to give. His heart pounds at the prospect. But he has a job to do. Her memory of the accident was only preliminary; there are a lot of basic questions her family wants answered. He has to do this.

He reaches out, closes his eyes, and touches her again. The most alien mind he has ever encountered. He remembers, with her, what she thought as he staggered away:

//We have to go back I have to go back please tell them to let me go back back back//

He remembered the white walls of the room. He remembered the hiss of the lights, the chair in front of him. He remembered his eyes seeking some avenue of escape, following every hairline crack in the walls to find somewhere he could dig in his fingernails and tear himself free. He remembered everything his body was doing in a state of visceral panic. His body knew they were coming to kill him.

His mind had not been there at all. He didn't know where it went or how long it was gone—but he knew what it was processing, and the synthesis the threat of death had prompted.

It was a manned mission into the past. He had returned from it whole.

The bolt was thrown back and the doorknob turned, and he was gasping "I have a proposal for you" as the Filipina came in with two armed men and a guy in a blue onesuit dangling a syringe from his fingertips as if it were a dead rat.

She took a deep breath: inhaling patience. She said, "Go ahead."

"If you do this with what you know now, all you'll have for your trouble is a melted spaceship and a dead pilot. You have to use the prefab gate, and no matter what telepaths you've lined up to cloud the minds of the Protectorate, their automated weaponry is still going to beat your telekinetics. If you want to waste Beth Atherton on a suicide mission, that's your business. But she's the only one who can navigate for you without a computer. Her Asperger's syndrome makes her a spatial savant. You have an alternative to a grand gesture of futility. But if you kill me, I'm not gonna tell you what it is."

"Thank you for the pilot recommendation. We'll make every effort to track her down." The woman was smiling again, and this time she looked genuinely amused—almost pleased. "If I torture you, you'll tell me the rest, and then I'll kill you anyway."

Jasper shook his head. "I don't think so. You folks really do mean well. Malcolm could be a bastard, but he was an idealist: he wouldn't have worked with you if he didn't believe in your good intentions. You're putting on a very good act to try to scare me, Pia Angelica, but I was there when you lost your baby. I was there when the baby was conceived. I know you better than you realize."

She blanched—clearly she had believed that Malcolm would never divulge those memories to anyone, though Jasper didn't think he meant to—but she didn't waver. "This project is more important than your life or his reputation. Don't underestimate that."

"I don't," he said quickly. "But if I'm on your side, and I can prove it with a gesture of goodwill, why not take me on?"

Pia glanced at the guy in the blue onesuit, who said he didn't have the time or the stomach for this anyway and slammed the door on his way out. Without dismissing the goons—very large scientists playing dumb, Jasper figured—she sat down and assumed a listening posture, elbow on knee, chin on palm.

"The pilots know that there's more than one way into hyperspace. One of them made a human egg of herself trying the alternative way, and she's dead now. I'm the only person who retains any part of her memories."

Deep breath. "Telling you that is my goodwill gesture. If you let me talk to your physicists, if you find Atherton, I can reconstruct it—both the theory the other pilots know, and the experimental data that particular pilot got firsthand. It'll tell you how to avoid Humpty Dumptying anyone else. She knew what she did wrong. She was ready to go back and do it right." It was all he had. He had made it as big as he could.

"We can't let you go back to your family," she said. "Not until the ship is under way and the news gets out on its own."

He blew air very softly, very shakily, out between his lips. "How long is that going to be?"

She ran a hand through her smooth bangs, the most human gesture he had seen her make. "Not long, I hope. I'm a little weary myself, Mr. Jasper."

"Yeah, threatening people's lives really takes it out of you."

To his relief, she laughed. "But Mr. Jasper, if we hadn't threatened your life, you would never have come up with your solution, would you?"

Joey's memories flooded out freely and joyfully, almost too many to grab and hold. Getting his first trike, the thrill of the wind, his pride in the colorful streamers flowing like captured water from the handlebars. Cuddling his big old bear as his mother sang him a lullaby. Dad teaching him to cast a fly into the trickling brown stream behind the complex, and the astonishing things they'd caught.

"Did I remember good, Daddy?" he said when they were done, when Jasper found himself once again looking into his son's sweet face instead of out from behind it.

This precious endowment of memories was the best reason he could think of for making sure he made it to Joey's adulthood. He knew what Katerina Thaïs knew, what Pia Angelica's crew now knew: there was a network of natural hyperspace conduits, like capillaries in the brain of the universe. They weren't wormholes as those were previously understood. They weren't measurable. They didn't leak matter. Particular pilots had the ability—an esper ability not unlike his own—to initiate them. Thaïs was one of those pilots, as was Beth Atherton: both autistic. Jasper felt sorry for the savants. He knew what it was like to have the guys in the lab coats after you.

But he was pretty sure that he also knew the reason for the Crash, now. What repeated immersions in the mercurial river of artificial hyperspace had done to the shipboard AIs. And when the refitted *Thorne* lifted, in two months' time, he was going to be on it, with Beth Atherton. Using the universe's own capillaries, they would slip past the blockade. They would go to Incarnatus, which the emissaries had never reached, because it, too, was under quarantine. They would hope that together they could find a cure for what the artificial wormholes had done to them, or deduce what caused the immunity the Protectorate was unaware of.

Or figure out if the Protectorate had done this *to* them.

The *Thorne*'s equivalent of an ancient aeronautical black box, he would bring that cure home, so that Joey might know in his own lifetime a world like the one the Crash had destroyed—or build a better one.

When he gave him back his childhood then, it would be a memory of steam and gaslight, the vapors of a lost dark age. He felt the vibrant current of Joe's young life, flowing so quickly into the old. He hugged the child to him against the rush of years, and said, "Yeah, Joey. You made the best memories I ever got from anybody, ever."

S. M. *Stirling* sold his first book in 1984 (*Snowbrother,* from Signet), and he became a full-time writer in 1988. That was the same year he was married to Janet Cathryn Stirling (née Moore), also a writer, whom he met at a World Fantasy convention in the mid-eighties. His works since then include the "Draka" alternative history trilogy (currently issued in a combined volume under the title *The Domination*) and the "Islander" series *(Island in the Sea of Time, Against the Tide of Years,* and *On the Oceans of Eternity)* from Roc Books, as well as *The Peshawar Lancers,* and a sequel to the "Terminator" movies (from HarperCollins). His most recent alternative history novel, *Conquistador,* was published by Roc in February 2003. He and Janet and the obligatory authorial cats currently live in New Mexico. Steve's hobbies include anthropology, archaeology, history in general, travel, cooking, and the martial arts.

THE CRYSTAL METHOD

S. M. Stirling

Ka-Rak had divested himself of his pack and cloak as well as most of his armor at the entrance to the wurm's cavern, and he was cold. Soon, he knew, battle would warm him—if the dragon's blazing breath didn't cure him of chills for all eternity. The warrior clutched the amulet that hung around his neck and kissed the leather bag that contained it to ward off the evil thought.

It was powerful magic; that he still lived was proof of that. But there were limits to what even the strongest talisman could do, especially against a wurm large enough to carry off cattle. Still, it was always prudent to offer respect to the spirits.

Ka-Rak looked around the rocky immensity of the cave and wondered what kind of spirits might linger here. The dragon had also taken a maiden and her younger brother; their angry souls might linger, crying out for revenge. He felt his scalp tighten as the hair on the back of his neck rose and shuddered.

Enough! he thought, and forced himself onward.

Spirits or not, he had taken this task on himself, and he would slay this creature or die trying. Dwelling on ghosts wouldn't help.

As he moved cautiously forward, the cave grew darker. He sniffed and smelled nothing. The absence of scent curled his brow; there should be . . . Suddenly the dry, sour smell of serpent filled the tunnel and grew stronger each moment. His puzzlement eased; clearly he'd been too distracted to notice, or perhaps had grown used to the stench and stopped smelling it until he'd turned his mind to it. Ka-Rak wished he hadn't done so; the smell was disturbing.

This tunnel must be a tight fit for the creature, Ka-Rak thought to distract himself. The ceiling was only four feet above his own six of height, though the tunnel was at least fourteen feet wide, most places; sometimes it narrowed. There were no loose boulders or stones on the rock floor and none projected from the ceiling. *Perhaps it is like a cat, and where its head can go its body can follow.*

Oh, *that* was a good thought! He forced his mind to a blank, totally aware state and advanced, his left fingers lightly touching the smooth rock.

The tunnel curved, and ahead was a glimmer of light, like firelight. Slowly, quietly, he drew his sword and shifted his shield from his back to his right hand as he crept forward steadily. As he moved, the smell changed to the sharp scent of overheated rock mixed with the rotten egg stench of sulfur. Ka-Rak adjusted his grip on the sword, listening with all his body and soul for movement, and heard nothing.

He kept his muscles loose, though the fear clawing at his belly wanted them tense, and he kept his breathing even, though with his heart galloping in his chest he wanted to gulp air. Once the battle was joined, fear would fall away and he would enter a timeless world composed of action and reaction, thrust and cut and parry. Even when fighting a beast such was the case: one had no time for fear.

Though in this case I might *be able to make room for it.*

The ancient tales had well prepared him to expect his greatest challenge. But also his greatest reward, for this dragon had a hoard. He grinned at the thought. Gold and gems beyond the counting, and all to be his, with the grateful thanks of the king and the hand of his lovely daughter.

Well, she isn't lovely, but for a princess she's not bad.

Some of the royal daughters he'd seen would send even this beast running.

The tunnel ended, and pressing himself against the rock wall, Ka-Rak carefully peered around the opening. The heat of the place struck him in the face like a blow.

The floor of the great cavern seemed to be a vast pile of embers, the dragon a dark shape lying upon them. The light they made was dim, barely illuminating the surrounding pillars of stone that held up a ceiling lost in blackness so complete it was as though nothing was there at all.

Immediately the warrior changed his strategy. A man couldn't fight in an oven, his feet would bake in minutes. And though he would love to trap the creature in the tunnel, there was nowhere to lurk in ambush, and he knew the beast was smart enough to precede itself with a bath of fire.

It will have to be in the open, then, he thought reluctantly. As if the wurm didn't have enough advantages.

It would have been better if he could have convinced men to join together with him in this task. But the peasants didn't have the heart for it, which is what came of a diet of beans and bread. And the nobles claimed it was dishonorable for any but one single hero to go up against the wurm.

Which was stupid, but then in Ka-Rak's experience, so were most nobles. If it wasn't for men like him coming down from the barbarian Northlands and giving them an infusion of new blood occasionally, the bastards would all have heads the size of apples.

He backed away, then made swiftly for the cavern entrance. Reclaiming his cloak and his pack, he went outside. Inside the pack were a set of hooks, a massive hammer, and a pile of spikes with which to hold the hooks in place. The noise he made setting the spikes around the entrance would draw out the dragon; like cats, they were curious creatures, and again like cats, they didn't like intrusions in their territory.

The purpose of the hooks wasn't to prevent the beast's escape, but to, hopefully, tear its wings to uselessness and prevent the wurm from taking to the air.

It had never been tried. Ka-Rak set to work.

* * *

Within its cavern the dragon slowly woke. Some vibration had penetrated its endless dreams of gold and fire, lifting it to consciousness. It blinked large, glowing eyes and yawned hugely, belching a small blue flame as it did so. Smacking its chops, the wurm looked about, seeking the source of disturbance.

There was a rhythmic pounding coming from the entrance to its sanctuary. Men would be the cause; no other creature would make such a noise. It yawned again, the plume of flame longer and brighter this time. It was not hungry, for it had eaten a cow before sleeping, and men were foul meat in any case—sort of greasy even when thoroughly cooked, with a sour aftertaste. But the intruder must be dealt with.

Men were a threat, and this one was making a direct challenge.

It stretched luxuriously, flexing its wings to their full extension, enjoying the freedom to do so. Then it shook its mighty head and eased forward, worming its way into the narrow tunnel, giving little puffs of flame, least the human be lingering nearby, saving its full blast for the entrance where the man might wait with a horse.

Horses were good eating. It hoped there would be a horse; even full it might like a few bites of the sweet, meaty flesh.

Bright daylight speared its eyes as it turned the corner and it blew out its breath in a ground-shuddering blast of fire, scuttling forward behind the safety of its shield of flame.

The fire was like nothing Ka-Rak had ever imagined, and it came with a roar that shook the ground and almost knocked him from his precarious perch above the cave entrance. The beast would come fast behind it, he knew, and the warrior struggled to stand firm. If the hooks worked, they might give him the precious seconds he needed to strike. If not, then the battle was most likely over as soon as the wurm could turn.

Gods, but the beast is fast!

Its head and most of its muscular neck were through before he could react. Ka-Rak dropped from ambush onto the dragon, landing just at the base of its neck above the shoulders. The wurm humped its back in reaction, breaking rock from the ceiling over most of its body and setting some of Ka-Rak's hooks deep into its shoulders. The beast screamed in fury and pain and turned its head to snap at the warrior.

Ka-Rak warded it off with the sharp point of his envenomed sword and clung tightly to its neck with his long, muscular legs. The point of his sword poked deep into a nostril, more by accident than design, and drew blood. The dragon drew back its head with a small sound that in a human would probably be "Ow!" It stared at its opponent.

In his experience, Ka-Rak had found that dragons couldn't speak and were no wiser than any other animal, including some humans. But they could hate, and needed no words to convey their loathing. Just now he could feel the malice in the wurm's glare, as palpable as its fiery breath. If it could draw itself out of the rock tunnel, it would roll, smashing him beneath itself. If it couldn't, it would back up and crush him against the ceiling. The dragon pulled against the hooks, bearing the pain they were causing because it wanted to be free to turn and kill.

Ka-Rak drew his dagger and, without taking his eyes from the dragon's, felt about with its point for a gap in the creature's scales. Like his sword, the dagger was poisoned, but it would take a hundred daggers like it to weaken the beast. Still, it would at least hurt more than being cut with just the dagger. When he found a place to slide the dagger in, he thrust into the wurm's neck with all his might.

White fire burst from the spot, and the dragon screamed in real pain. Ka-Rak threw himself to the side and almost fell from his place. The dragon pulled itself forward, yanking the hooks from their moorings. It shook its mighty head from side to side, roaring in fury and agony. The sound was like a blow, hurting his ears terribly. The warrior clung to the beast's neck with one muscular arm and hacked at the burning place with his sword, trying to widen the wound. Hard going through its protective scales and the wurm's plunging motions, but he was making progress. Flame flared and dripped from the wound.

Then the beast was free of the tunnel. Ka-Rak leapt from its back and rolled, coming up hard against a boulder. Yet he was on his feet instantly, crouched and balanced to leap to either side, his eyes checking the ground for advantage, while the bulk of his attention stayed on the dragon.

It was checking its wound, almost, but not quite, touching it with its clawed hand. It put its nose close to the spurting fire and touched it with its long snake's tongue. It licked up a drop of fire, then shook its head, snorting. Apparently it didn't taste very good.

His shield was on the other side of the dragon, meaning his only protection should the dragon decide to shoot flame at him was the boulder he'd come up against. Ka-Rak slipped behind it and assessed the dragon's other wounds. The hooks had damaged the creature's wings, how badly he couldn't tell, but some of the hooks were still embedded in the wurm's flesh. At the very least they would be painful, and the loss of blood would be a help to him.

As if it had heard his thought, the dragon's head snapped around to glare at him. It crept forward with the careful grace of a hunting cat, its head weaving back and forth on its sinuous neck. The creature met Ka-Rak's eyes and almost seemed to grin when the warrior divined its intent. Then the beast reared back.

Screaming his war cry, Ka-Rak leapt over the boulder and thrust his sword forward like a spear. To his surprise, it sank deep into the base of the creature's throat and stuck. He struggled to pull it out for a moment, then, wincing, glanced up. The dragon looked straight down at him and seemed to sneer; then it drew in a long, deep breath, preparing to blast him to oblivion with its flaming breath.

Suddenly Ka-Rak was flying, his gauntleted hands on fire. When he struck the earth, though the breath was knocked from his body, Ka-Rak concentrated on pulling the flaming gloves from his hands. *Then* he set about writhing on the ground, trying to breathe, simultaneously putting out the minor fires that burned on his hose and shirt. Eventually, gasping and feeling as though he'd been beaten with clubs, he turned toward his opponent, his wounded hands clutched to his chest.

The dragon's headless body was aflame, its ruined wings twitching feebly—the cooking meat smelling oddly like roasting chicken. Several feet away in a twisted heap lay the beast's head and a good length of its neck.

Ka-Rak smiled, thus discovering that his face was burned also. Still, relief coursed through his body like a strong wine, leaving him giddy.

Looks like there's going to be a royal wedding after all, he thought, and laughed . . .

. . . and Ken Rackam checked his weapon—less than a quarter charge left—with a mild annoyed oath.

With four of Kletzer's goons after him, he might as well just use it on himself and save everyone some trouble. He put the weapon back in his pocket and pressed the seal shut. This assignment had been screwed up from the start.

When he'd left Discrete Couriers, Rackam had briskly walked a mile in a random direction, window shopping here and there, then stopped for coffee at the first available place. It was a routine of his to establish that he wasn't being followed. In this case, it established that he was. That pointed to a leak in the company, a definite complication. Especially for a company whose reputation depended on its discretion.

His tail was a medium everything kind of guy, Joe Average from his shoes to his haircut, and he was pretty good at his job. The thing was, Rackam was better. The guy at least had the sense not to come inside and order a cup of coffee, but after ten minutes it was hard for him to blend in. People don't just walk up to a shop-front and fall in love with it without going inside. And unfortunately for his tail, this neighborhood seemed to specialize in maternity fashions, baby gear, and toys. There wasn't even a corner news station he could pretend to be fascinated by. Rackam could almost feel sorry for the jerk.

He paid for his coffee and headed for the unisex washroom, where he locked the door and with his weapon cut himself an opening in the plasticrete wall that separated the coffee shop from the store behind it. He found himself in a deserted storeroom and carefully lifted the plasticrete slab, shoving it back into place. With luck, the edges were still hot enough to stick together, making yet another obstacle for his tail to get through.

He walked to the door and found it locked. The courier rolled his eyes and swore in exasperation. Opening the seal on his breast pocket, he took out a universal key and swiped it across the lock mechanism. The door clicked open. Before going through, Rackam reversed the key and swiped the other side across. This would freeze the lock, leaving his tail the choice of blasting the thing off or going back through the wall and around the block. Either way would be very inconvenient.

Rackam allowed himself a small, evil smile; it would probably put the poor fellow in quite a temper.

He'd been ordered not to contact Discrete until he'd completed his mission. An unusual request, but not completely out of line, particularly for a high-level job. Now he wondered. No conditions had been put on the restriction, which meant there were no exceptions. So, he was without resources beyond his native wit and whatever he carried on his person until he finished his assignment. And, as ever with Discrete Couriers, failure was not an option.

It should have been simple. Go to station quadrant A, level fifteen, platinum section, clinic 17, see Dr. Ho. There he'd give a sample of his blood, encoded with a biomessage, assignee unknown, receive the antidote that would clear the foreign DNA chain from his bloodstream, and his part of this nonsense was finished.

Instead he was being followed. No problem. As usual he would avoid his destination until he was certain he was unobserved. He checked his watch; he had leeway. Rackam left the electronics store, to the surprise of the employee who hadn't seen him come in, and started his second random mile.

This time he walked faster, turned more corners and twice doubled back. The second time he'd planted a small, self-sticking recording device on the side of a shop. When he retrieved it and played it back, he swore. He was still being followed, if not by the same guy, then by his twin.

Rackam didn't like this, not one bit. He still had time before he was due at the clinic, but it wasn't endless, and this guy was good. Or at least better than the courier had first thought. He kept walking. Time to try something else. He swung into the first public transport station he came to and waited.

When the transport came he waited until the last moment and entered the most crowded car. At the next station he changed; at the next he entered a transport through one door and dashed out another at the last second. Then he stood by a news station; he palmed the recording device and mimed brushing his hair back. Then he replayed the scene behind him.

Good, no sign of the tail.

For the first time Rackam looked around to see where he'd landed and grimaced. Quadrant D, level six, gray section, aka—the Dark Zone. Not a good neighborhood. Many of the lights were out—due to vandals, no doubt. This was true throughout the area, just one of the reasons for its nickname. The transports and the crowds, which had been so cooperative until now, seemed to have disappeared.

Another inconvenient little quirk of the place? He frowned; the odds of getting a personal transport in this area were pretty near zilch.

Failure is not an option, he reminded himself, and crossed the platform to wait for a transport back to the previous station. That one was a transfer point, and he could get to a higher station level and a lot closer to platinum than he currently was.

The Dark Zone was a tolerated aberration. Meaning that the psychs had found that people, especially young men, seemed to need a touch of the illicit to function at a high level. Hence this section of the station where almost no one lived, but certain businesses thrived, certain types of entertainment and highly restricted, or even outright illegal, drugs were available. Stationers came here in disguise to cut loose, defy authority, and escape the drudgery of their everyday existence.

It was the only place on the station you could get mugged. You wouldn't get murdered—the station still maintained security cameras here and would shut down anything that looked like it was getting out of hand instantly. But you could get the crap kicked out of you in the meantime. *All part of the fun,* Rackam thought.

Sometimes he wondered about the messages he carried. This one must be a humdinger, because getting to someone inside Discrete would have to be expensive. The penalty for betraying customer confidence was a crushing fine, a year in jail and exile from the station. Very unattractive prospects. Especially since your transport home would be at your own expense. Essentially you'd be a slave to Discrete for the rest of your miserable, poverty-stricken existence. *Why would anyone risk it?*

There was a footfall, and he slanted his eyes in its direction, subtly adjusting his posture to a more defensive state. More footsteps came from the opposite direction, then more and more until he sensed he was surrounded. Something in him wilted in exasperation. What were the odds this was just a bunch of people waiting for a transport?

"Hey, stinch," a voice whined from the direction of the first set of footsteps, "can ya spare me a credit?"

After which me and my friends will dance on your head, the courier mentally continued in the same nasal bleat. Rackam put his hand in his hip pocket and clasped his weapon.

"No," he said calmly. He had the weapon out and aimed before the would-be mugger could react. "But I can blow your kneecaps off."

The would-be thief held up his hands with a nervous laugh. "Hey, stinch, no need to get mad, y'know? I's just askin'." He backed off.

"Tell your stinches to take a hike, too," Rackam advised.

"Yeah, sure. Hey, guys, let's go." He jabbed a thumb over his shoulder.

Others came out of the shadows and trooped toward their leader, snickering at the courier as they passed him. Suddenly one of the thugs snapped a fist at Rackam's head. The courier caught the hand and viciously twisted it high up behind his attacker's back. Before the man's shriek had a chance to echo, Rackam had his weapon trained on the gang leader.

"Did I say kneecaps?" he asked politely. "I must have meant head."

"No, no, no, no," the leader said so fast it sounded like he was stuttering. "You don't wanna do that. Just give him to me, and I'll take care of this. I din't know he was gonna do that. Just keep it zen, okay?"

Rackam made a move that sent the thug rolling to his leader's feet, where he was greeted by a kick to the head. Almost immediately the others came from the shadows to add their boots to the fray while their former stinch curled into himself protectively.

The courier looked at them in disgust. People could be such animals. Undoubtedly these idiots had jobs that required them to be courteous and cooperative for eight hours a shift. Yet here they were acting like hyenas. *Probably that's an insult to hyenas,* Rackam admonished himself.

He heard a transport coming and didn't want to take his eyes off the group of thugs to see if it was the one he was waiting for. He fired a warning shot, and the whole group hunched.

"Take it somewhere else," Rackam snapped.

"Sure," the leader said. He snapped his fingers, and his boys started drifting off. The one on the ground struggled to get up and failed. The leader snapped his fingers again, and two of the thugs hoisted their gang-mate to his feet and dragged him off. The last to leave was the leader. He pointed a finger at the courier.

"I'll remember you," he said, eyes narrowed.

"I can't say I'll return the favor," Rackam replied. "You'll be just as insignificant the next time we meet as you are right now." He raised the weapon in an indication that it was time for the mugger to join his friends.

The man hawked and spat, then, with a last glare over his shoulder as he turned, walked swiftly into the gloom, his coat swinging around him.

Nice exit, Rackam thought. The courier hoped he never met this crowd unarmed; guys like them held grudges.

A transport rolled in across from where he stood, and through a window he spied his average guy tail. Rackam swore and brought his hand up to his ear as though stroking his face. Too late, he knew, the tail had been looking right at him as the transport swept to a stop. Maybe he should just wait for the guy to stroll over and introduce himself.

But when the transport rolled away, there was no sign of the man. *Just when you think you've got the whole thing figured out, they go and change the rules,* Rackam thought.

"Mr. Rackam," a voice said from behind him.

The courier snapped his head around to find himself confronting his tail. Only he couldn't be the tail because he was dressed completely differently from the man who'd just arrived on the transport. Instead of tan, this man wore black; instead of casual, this man was dressed in a business suit.

"What are you, a clone?" Rackam asked.

"Cloning is illegal, Mr. Rackam," the man said. "You may call me Leon."

"What do you want, Leon?"

"Just a sample of your blood and an opportunity to talk you into allowing us to make a slight change to the message you carry." Leon stood calmly, with his hands clasped in front of him as though he'd merely asked for a moment of Rackam's time.

"Oh, that's a nonthreatening request," Rackam said.

There was a sound off to the side, and he side skipped several paces away, lifting the weapon to keep both Leon and whoever approached in his line of fire. It was another Leon. From the clothes, this one was the one from the transport. This one looked from Rackam to Leon.

"Kletzer told us to explain," he said.

"Kletzer?" Rackam asked.

"The one who told us to intercept you," Leon clarified.

At that moment a transport swept up behind the courier, and one of the Leons turned away. Rackam hid his weapon but kept it in his hand. Leon started forward as the courier backed into the transport but stopped when Rackam shook his head. Leon dropped his hand, looking disappointed. The doors closed, and the transport took off. It wasn't until then that Rackam thought to look for Leon #2 and swore when he didn't see him on the platform.

Kletzer, he thought. Who was Kletzer? The name was vaguely familiar, but he couldn't place it. Looking around, he quickly established that Leon #2 wasn't in this car. That left Leon #3 unaccounted for. The other man was also casually dressed and wore tan, but the cut of his clothes was slightly different. He checked his weapon—a little more than a quarter of a charge left. Cutting through the plasticrete had depleted its charge considerably.

Time to take a more proactive approach. As the transport slowed to a stop, he positioned himself in the doorway. He left with a crowd, walked forward for a few steps, then started walking backwards until he was back inside, to the considerable annoyance of his fellow passengers. From the corner of his eye he saw Leon #2 step back into the car. Rackam waited a few seconds, then stepped out again, standing on the platform until his tail came out again, then he stepped backwards into the car. Once there, he waited until he door began to close, then slipped through the gap simultaneously snapping off a quick shot toward Leon's door. Leon hopped backwards, losing his chance. As the transport pulled out, Rackam could see his lips moving, signifying that he was making a call to someone. Hopefully Leon #1 back in the Dark Zone.

The courier hurried to an elevator, catching it just as the doors closed. If he was lucky, the computers missed that shot of his; if not, he'd have some explaining to do when the elevator doors opened.

Gradually the elevator emptied as they passed the working-class floors until he stood alone at the front with his back to the control panel. Soon he'd have to use the pass key that would allow him access to the platinum section. For now he could relax and watch the advertisements play on the sides of the cab.

The doors opened. Nothing happened for a moment; then a man hurtled into the elevator, taking up a position with his back against the far wall, a dart gun in his hand. Rackam kicked out, and the gun went flying. Leon #4—by his clothing this was yet another one—kicked back, and the courier barely managed to sidestep the attack. Cloning was illegal; it seemed to be far too common, though.

Fighting in a confined space like this was unpleasant. Leon #4 rammed his fingers toward Rackam's eyes; Rackham snapped his head aside, slapped his right wrist onto Leon #4's and used momentum and a twist of his hips to sail the clone into the elevator's thin metal wall. His face made an ugly splatting sound; Rackham pushed off in a twirl as he released the other man, pivoting on one foot with the other coming around like a scythe—heel-first. It punched into the clone's body just below the ribs with an impact that jarred up Rackham's body and into the small of his back. An unpleasant sensation, but much more unpleasant at the other end of the kick.

Things got easier after that. When the doors opened again on the last public floor Rackam grabbed Leon by his collar and pants and flung him into the corridor. Leon #4 lay panting for a moment, then gamely struggled to his feet. He was staggering toward the door when the courier raised his weapon and Leon #4 stopped. He stared at Rackam and shook his head.

"We need to talk to you," he said.

Rackam didn't answer, just kept his weapon trained on Leon's middle and his eyes on Leon's bloody face. The doors closed, and Rackam folded over, his hands on his knees and groaned. He felt sick, and he had no doubt that he looked like he felt.

"In order to proceed," the elevator's smooth feminine voice prompted, "it will be necessary for you to insert a pass card into the slot above the floor selection numbers. If you do not insert a pass in ten seconds, this car will be going down."

He inserted the pass key, then checked his weapon; *less* than a quarter charge left and *four* Leons after him. They *had* to be clones. For the first time he seriously began to doubt his ability to complete his mission. The doors opened, and he was in platinum sector.

A small dark-haired woman with a thin face stepped into the cab just as he tried to exit, and he stepped back instinctively. Rackam barely felt the prick of the dart before he hit the floor. He looked up at her, unable to move. It wasn't fair!

He'd been expecting a Leon.

As he woke, he found he'd been bound with tape, but not gagged. He heard people talking, and gradually their words began to make sense.

"You shouldn't be here!" a man insisted. It sounded like Leon.

"But I am here," a woman said, her voice sounded weary. "He needs to know. He has a right to know."

Silence greeted this remark, so Rackam broke it.

"Know what?" His voice rasped in his own ears; he wondered how long he'd been out.

The woman came and stood over him. Suddenly his mind cleared, and he remembered who she was. A tech from Discrete that he knew as Carolyn.

"Kletzer, I presume?" he said.

She smiled slightly and nodded.

"Why?" he asked.

"Because you're as human as I am," she said. "But not legally human. You're a genetically altered clone slave." She gestured behind her, and he turned his head to see the Leons in a semicircle, their identical faces serious. "Just as these men were."

The Leons nodded.

"What they never tell you is that to transport information in a DNA strand is illegal if the bearer is human. That's why you exist. They took material from a dedicated courier and cloned it, fast growing it to maturity after taking out and adding certain attributes. Your appearance is slightly different from the original, and you've been deliberately given a low-affect character. Something that was encouraged by the programming they fed you as you grew. When you've completed an assignment, you're put into stasis until you're needed again. You also have three brothers, none completely identical to you."

"Stasis?" he rasped.

Rackam licked his lips, and Carolyn—he couldn't think of her as Kletzer—offered him water in a zero-g bulb with a straw attachment. When he'd taken a few sips, he turned his head away.

Carolyn sighed. "Essentially you're in hibernation, your metabolism is greatly slowed, and you're given special antiaging agents they'd never dare administer to a legal human. They feed you programming that keeps you updated and creates the illusion that you have a life." She shook her head sadly. "But, Ken, you don't. It's all a lie. Dorrana is a construct," she said, referring to his girlfriend.

Of all the things she'd said to him, that was the most disturbing. Perhaps because there *was* something about Dorrana that wasn't right. She was too compliant, too available, entirely too perfect for him. It was the most convincing thing Kletzer had said so far. He put it aside. There were more important questions to ask.

"Why now? Why this message?" he asked.

"What you were carrying was a report outlining the possibilities inherent in legalizing and even extending the clone program," Carolyn explained. "It was directed to the entire upper council of the station, and the punch line was a stock offering that would make them rich beyond their wildest dreams of avarice in spite of the trouble it would cause between the station and Earth. There was also a strong subliminal command, the ultimate in an ongoing program of brainwashing that would convince them utterly of the rightness of the proposal."

She clasped his shoulder. "Don't you see, Ken? Thousands of people just like you would be created and enslaved, without even the comforts and privileges that you've been given. With even less choice, and far more dangerous and arduous labor to look forward to until they're worn out and replaced like machine parts. People with a built-in expiration date."

Carolyn raised her hands and dropped them again. "Can you see why I had to act?"

Rackam closed his eyes and thought about what she'd told him. Could she possibly be telling the truth? Slaves? *He* was a slave? Somehow he seemed to want to believe her. Was she using some kind of mind-control drug on him?

Forcing calm on himself, he examined his life, thought about his childhood, of which he had few memories. Then he recalled the occasions when he'd wakened, briefly, in some sort of isolation tank, floating in water that must have matched his body temperature because he could hardly feel it. Before he could draw a full breath, he'd be back in his bed, or eating a sandwich or whatever he'd been doing before the strange vision.

A mild alarm stirred within. Now that he thought of it, he realized that it had happened numerous times. He'd thought it was just a recurring nightmare, but now . . . Carolyn's explanation matched the experience far more than his reality did. His life *did* move in fits and starts. Things changed very suddenly for him, surely more suddenly than could be normal.

"She's telling the truth," Leon said.

Rackam nodded. "Yes," he said aloud. "It answers all sorts of questions." He opened his eyes. "What do you want me to do?"

Carolyn and the Leons visibly relaxed. So they did need something from him.

"We cleared you of the other message," she said. She held up a hypo. "This contains a new message, demanding a stop to the illegal cloning and freedom for the clone slaves. Pointing out that any other action could lead to war with Earth, which takes a very dim view of slavery. It leaves the subliminal command to obey intact."

She bit her lip. "I know that's wrong," Carolyn admitted. She shook her head sadly. "It makes them slaves, too, in a way. But if we don't do it this way, then there'll just be another message that will supersede any amount of reasonable persuasion. It has to be this way or they'll win. Do you understand?"

The courier nodded. He understood. He also believed that if it were just a matter of convincing the council to do the right thing, Carolyn would have gone to the wall over it. Come to think of it, she already had.

"I'll do it," he said before he could change his mind. "How much time do I have left?"

"Fourteen minutes," she said.

Rackam blinked. They must have given him something to wake him up.

"Do it," he said.

Carolyn pressed the hypo to his arm, and he felt the shot go home. By the time he reached the clinic, the message should have spread through his blood to the point where Dr. Ho wouldn't notice any lack of material.

One of the Leons sliced the tape off and helped him to his feet.

"Everything has to look normal," he said. "That means you've got to go back to Discrete."

Rackam nodded, his face grim.

"But if this works, you'll be free soon," Carolyn reminded him. "Now, hurry. Even though you've got time left on the clock, you've taken an unusually long time getting there. It's important that they not become suspicious." At his nod she offered him her hand. "Good luck," she said.

He shook hands with her and smiled slowly. "Don't worry, Kletzer. Failure is not an option."

* * *

In a darkened room subtly scented with lavender, a gentleman lay upon a couch upholstered in a lavish silk brocade. Embroidered pillows cushioned his noble head, upon which a cunningly crafted device constructed of copper had been placed. It held three glowing crystal wands with their points just touching his forehead. His eyes beneath their closed lids swept back and forth frantically as though trying to see myriad images flashing by at speed.

The light slowly faded from the crystals, and the gentleman's eyes gradually became still. For a few moments he slept. Then, as the room became gradually brighter, his eyelids fluttered open and he stretched. Sitting up with a quick, decisive movement he glanced around the room, then rose and went to a dressing table, upon which his elaborately curled wig rested on a stand.

He sat and removed the device from his forehead, examining it carefully until he was interrupted by a knock on the door. The gentleman looked in the mirror and raised one brow in mild irritation.

"Come," he said in a languid voice that belied his excited mood.

"Good evening, m'lord," the wizard said. He was an elderly man in a long spangled robe, and he held out his hands for the instrument that the nobleman was examining.

Reluctantly that worthy returned it.

"A most entertaining afternoon, good wizard," he said.

"I am glad Your Grace is pleased," the wizard simpered.

The duke swung the heavy wig onto his close-shaven head and adjusted it. Then he rose and put on his embroidered and spangled coat, twitching the exquisite lace of his cuffs free of the sleeves.

"Indeed I am," the duke replied, making a final adjustment. "I am quite certain that His Majesty will enjoy this new amusement, as well."

He smiled thinly at the sorcerer, conveying with his expression that the wizard had better be grateful for this opportunity to pander to the royal whim.

No fool he, the wizard bowed low. "I shall be forever in your debt for bringing my humble efforts to His Majesty's attention," he assured the duke.

Picking up his gold-capped ebony walking stick and adjusting the set of his rapier, the duke proceeded to the door, where he turned to stare at the wizard. After a moment of this scrutiny, the wizard began to fidget.

"Is there some way I may serve Your Grace?" he inquired.

"I was just wondering . . . where *do* you get your ideas? All the other entertainers of your ilk draw *theirs* from ancient legends or ballads. Yours—"

"Ah, Your Grace," the wizard said, bowing. "Mine are drawn from out of the very ether!"

The duke's expression turned haughty. "Hmmph!" he uttered, and left without another word.

The wizard wiped sweat from his brow and placed the copper device on the wig stand, then exhaled a long breath. He chuckled delightedly and rubbed his hands together in sheer delight. His fortune was made! How jealous his fellow wizards were going to be! And all with so little effort on his part, too. He foresaw honors and gold showing down upon him; life was good!

Turning, he swept a tapestry curtain aside to reveal a pane of glass set in an elaborate copper frame. Behind the glass streams of smoke in many bright colors swirled and billowed, yet never mingled, as though tightly contained, yet blown by a continuous wind.

With great care he removed the crystals from the headpiece and placed them in similar holders attached to the rune-scribed frame. As soon as the third crystal was in place within the frame, the colored smokes began to swirl and clear away. A quacking voice blatted, "Sci-fi!" and before his eyes four people in loose, mottled clothing walked toward a great upright circle of stone containing what appeared to be a pool of water, impossible given its position. One by one they disappeared beneath its rippled surface.

The wizard drew up a chair and sat down, cackling with glee. He loved this program! It was even better than the one with the tall dark-haired warrior woman and her blond friend. . . .

Alex Irvine is the author of *A Scattering of Jades,* (Tor Books, 2002) and the forthcoming *One King, One Soldier* (Del Rey, 2004). His short story collection *Unintended Consequences* is due from Subterranean Press this year. He has also published short fiction in (among other places) *F&SF, Asimov's, Sci Fiction,* and *Lady Churchill's Rosebud Wristlet,* and was coeditor of *The Journal of Pulse-Pounding Narratives.* He lives in Portland, Maine. His Web site is http://alexirvine.net.

REFORMATION

Alex Irvine

ALIF LAM MIM

Enter.

ALIF LAM MIM SAD

Marwan does one last dry run before putting his life on the line. He slips through the outer layers of Southern Baptist Convention security, in and out like a needle through a balloon, and then the Assemblies of God. He does not bother with the Lutherans or Presbyterians or Methodists; they have no security worth noting, not really. The Methodists even permit their congregants to own personal computers. The great temple in Jerusalem is more difficult, especially since the virus terrorism of 2013, and the thickets of security protecting the magnificent Mosque of the Prophet in Medina are a still sterner test of his abilities. Marwan does not dare approach the Vatican, not yet. He will get only one opportunity there, and he will have to make that one count.

Ghosting in and out, he monitors his pursuit. A few strays from the Assemblies of God, a few more from Jerusalem and Mecca. All easily dusted off. The kind of thing he's been doing since he was eleven.

He sits in his apartment off Ford Road in Dearborn, Michigan, with the shades drawn. It is five in the morning. Time to pray. Marwan removes his sneakers and his Detroit Red Wings baseball cap. He knows he should wash, but he does not. He unrolls his prayer rug in the living room and kneels on it, facing the balcony. As he bows to touch his forehead to the rug, the first rays of the sun find their way through the drapes and strike the top of his head.

ALIF LAM RA

Marwan never prays to be saved, or to be martyred. He never prays for the souls of his mother and father, dead in an auto accident the previous spring. He never prays for the money to go to university; he can read, and his mother taught him how to learn. He prays only that he is doing the right thing, and asks God to tell him if he is not. So far he has not heard from God, and he carries with him a bone-deep certainty that he has understood the signs of the times and that he has chosen the only course of action possible.

He speaks of this with no one. Not the rental office on the first floor of his building, where he drops off the check on the first of every month. Not the restaurant where he buses tables and washes dishes to afford his three rooms. Not the four hundred or so other worshipers at the Beit Jalal mosque in Dearborn Heights.

And especially not those faceless searchers he encounters on-line, his fellow travelers. He is more careful with them than with any church or synagogue or mosque.

Not for long, though. Fairly soon all his cards will be on the table.

The sun is fully up, and the day promises to be hot. Marwan opens the drapes and steps out onto his nineteenth-floor balcony. A couple of miles from his building, Dearborn gives way to Detroit: a lesson in the infliction of the sins of the father onto the sons. The assembly line took its first steps in Dearborn and sowed the seeds of its own destruction in Detroit. As Marwan gazes off to the east, the sun shines down on 900,000 Detroiters in a city built for two million. From where he stands, Marwan can see trees growing through the roof of an abandoned factory that once produced ball bearings.

He cannot help it; this is what he sees when he thinks of the Neoplatonists—Augustine, al-Farabi, and all the rest—and their City of God. Henry Ford had been Detroit's tin god, and the results spread ruined and empty as far as Marwan can see. Down near the river, heavy traffic congeals around office towers, buses idle in front of casinos, new stadia shine brightly in the June morning. But between Foxtown and Telegraph Road, the city is as empty as if a great plague had passed.

Will I change any of that? Marwan asks himself. And has no answer.

ALIF LAM MIM RA

Of all the numbers, the number nineteen is the most holy. The number of verses in the Qur'an is 6,346—a multiple of nineteen, whose digits add up to nineteen, as well. The number of suras in the Qur'an is 114, a multiple of nineteen. The *basmala* occurs 114 times in the Qur'an, and each *basmala* is nineteen letters. The first and last revelations consist of nineteen words. And so on. Nineteen, Marwan believes, is the number of God. This is certainly true in Detroit, where the number nineteen hangs in the rafters of Ilitch Arena, memorializing the great Steve Yzerman's four Stanley Cup triumphs between 1997 and 2003. Marwan is a great fan of the Detroit Red Wings, and when he sees that number *19* hanging above the ice, a peace comes over him. Surely he has chosen the right place.

He is back at his computer, preparing for the day. A distracted part of his mind wonders what Martin Luther thought of the number nineteen. Another distracted fragment wonders if he might better have pursued his goals in Denver, where Joe Sakic wore *19,* or New York, where Mike Bossy wore *19,* or . . .

Names from my childhood, Marwan thinks. Names my father would mention when we were watching the Red Wings. *"Poniatowski reminds me of Mike Bossy, you know, he wore nineteen, too. But Yzerman was the best of all of them."* And they have percolated into my mind, become my own comparisons and touchstones even as I put away childish things.

In high school, Marwan read a play called *Everyman*. Summoned by death, Everyman fears that he cannot complete the voyage to Heaven alone. His friends Kindred, Cousin, Fellowship, and Goods cannot accompany him. In the end, only Good Deeds—previously neglected—steps forward and finishes the journey with Everyman. Along the way, Knowledge offers himself: "Everyman, I will go with thee and be thy guide." The line has not left Marwan's head since he read the play, now eighteen months ago.

Marwan's father and mother were still alive when he read the play, still there with him to be his guides. No state policeman had yet come to his door with rounded shoulders and shifting eyes. No doctors had yet touched him on the shoulder and left him alone in a cold white room with two sheeted bodies. And he had not yet suffered the comfort of hearing that all was well, that Imad and Ayat Aziz were gone to whence they'd come, absorbed back into the divine. Gone home.

He had gone to school the next day because he didn't know what else to do, and he stayed after Edsel Ford High School had emptied of everyone but custodians and off-season football players. Not knowing where else to go, he had wandered into the shop classroom, where old Mr. Krause was still tinkering.

"Krause," Marwan said.

Krause was the only person in the world Marwan knew who seemed like he really knew what was going on. You were having trouble with an English paper, your car, your girlfriend, you went to Krause, and he made it all make sense. He was about five feet seven and weighed maybe three hundred pounds, wore his hair slicked back over his bald spot, and collected kids' toys from the 1970s. A strange guy; the general consensus around Edsel Ford was that the cops would come looking for him one day.

But on this day, it was important to Marwan only that Krause knew what was happening inside him.

"Z," said Krause. He had a nickname for everyone.

"My parents were killed yesterday."

"I heard." Krause looked closely at Marwan. "It hasn't hit you yet, has it? You're still—"

"Are they gone, Krause?"

A pause. "Depends on how you mean that, I guess, but I'd have to say yeah. They're gone."

"Not in paradise, or heaven, or nirvana, or any of that shit."

"Z, are you a religious kid?"

Marwan had to think carefully about this. "I think so."

"That's the problem with this, isn't it? Someone told you something, and you don't know what to do with it."

"One of the nurses," Marwan said slowly. "She said my mom and dad were gone home. Taken back into the divine or whatever."

"Reabsorbed, huh? That's what's bugging you."

"If that's what happens, then they're dead in heaven, too, Krause. What's the point?"

Krause shifted his bulk in his chair and stroked his goatee. Marwan began to cry. The tears came slow and hard, and each breath tore itself loose of his diaphragm. Krause let him go for a while. Then he said, "Z."

Marwan wiped his eyes.

"What the nurse gave you was emanationism. All worldly things come from the divine and disappear back into it when they die, and so on. Neoplatonism, the whole idea of forms and the material world being a debased reflection of what's up there." Krause pointed to the water-stained acoustic tiles over his desk. "You, you probably grew up thinking you were Muslim, even though any kid who really grows up in America can't have a pure religion. It's not that kind of a place. Here's the test: If you're a real Muslim, you have to believe that the Qur'an was dictated to Mohamed by God, and that even though it's full of contradictions and long bits of just plain loopiness, it's the infallible word of God. You believe that?"

No, Marwan realized. I don't. I never did, really.

In that moment something fell into place within him, and he fell away from the comfortable osmotic religion that had been the background noise of his childhood and adolescence. Krause saw it happen.

"Right," he said. "That's crazy. No rational person can believe that. Mohamed was a reformer, Z. He looked around him and saw all the craziness on the Arabian peninsula fourteen hundred years ago, the tribal wars and infighting, and then he saw how the Christians were going gangbusters up north and the Jews were holding on like they always had, and he had a divine vision that would make it all right. For my money, Mohamed was more like Martin Luther than like Abraham. Luther nailed his theses to the door because he wanted to get the church back to the source, the unmediated relationship with God. And that's the argument of Islam, that the other People of the Book have gotten it all wrong, and that the Qur'an will come along and get back to God." Krause shrugged. "You should go off and read the Sufis. You don't believe the Qur'an is divine dictation, you're a Sufi anyway. They're all about looking behind the words, them and the Kabbalists and the rest." He reached out a meaty hand and punched Marwan in the shoulder. "You've got a lot to think about. I'm sorry about your mom and dad, Z. Come back if you want to talk more."

"Yeah," Marwan said.

"Oh. You're into math, right?"

Marwan nodded.

"Then you should check out the Brethren of Purity, Z. They'll be right up your alley."

Marwan went straight from Krause's classroom to his car, and then straight out I-94 to Ann Arbor. No way would the Dearborn library have what he was looking for. He needed a university library. He found the Ikhwan al-Safa, and the Sufis, and read until the library closed at midnight, and that night he went home and in the midst of his mourning thought, All over the world there are people like me, suffering because they are told what is not true. Somewhere my mother and father stand together with God, and are complete. But who has gone with them to be their guide?

Mind wandering still, Marwan thinks that Martin Luther and Mohamed would have had much to talk about. A grin passes over Marwan's face as he wonders if the Prophet ever dispelled the Devil with flatulence, as Martin Luther so famously did.

Today is Marwan's nineteenth birthday.

ALIF LAM RA

He has planned his action for early afternoon, when network traffic is at its most intense: East Coast American markets closing and people checking e-mail before leaving work, business use on the West Coast peaking, and markets in the Pacific Rim heating up along with the business of the day in Tokyo, Shanghai, Bangkok, Kuala Lumpur. It will be 1900 Greenwich Mean Time.

He takes a disk from an envelope and watches the sunlight play across its shining surface. He has created this on a computer far from his home, at public terminals in Walled Lake and Eastpointe and Inkster and even Windsor, south across the river. All his reading, all his desire, all his sorrow and hope have gone into it. Somewhere in its rainbowed interior, the broken symmetry of the technical and the mystical is restored.

The disk rests on the lip of a slot in Marwan's terminal. The tip of his right index finger exerts gentle pressure on the rim of the disk. There is no way to be certain that what he is planning will work. It may be that when he inserts the disk, he will wreak changes greater than he imagines.

What changes?

God: stop me if I am doing wrong.

He does not know which God he expects to hear his prayer. With a soft whir the disk disappears into the slot.

KHAF HA' YA 'AIN SAD

Marwan has named the font Brethren, after the Ikhwan al-Safa, the Brethren of Purity. The Brethren had flourished briefly in Basra nearly a thousand years ago and left behind only one work, a cryptic and self-contradictory collection of notes amassed over a period of years. Reading the Brethren the year before, with his parents still unburied and a great many of his tears still unshed, Marwan had thought, This is crap. Emanationist crap, to borrow Krause's word. Why did he want me to read this?

Then a diamond, the kind of signal idea that redeems hours and days spent bleary-eyed in the stacks of the university library. The Brethren had believed that the study of Number—of astronomy, geometry, algebra, and music—would bring the student to the mind of God. The simplicity of this idea took Marwan like a lover, and he clung to it as he read Pythagoras and even when he read the attack of the great al-Ghazali, who excoriated what he saw as the Brethren's watery ecumenical mysticism, claiming they denied the omnipotence of God.

But the Brethren had seen no conflict between philosophical and religious truth: Mathematics, number, was the way to the mind of God.

Computation on the fingers, making the body into a number, approaches divinity. (And did not the Brethren and al-Farabi and even al-Ghazali—and the Christian Gnostics—agree that when a man knows himself he begins to know God?) Computation in the gold and silicon universe of the microchip is as near to God as man has yet approached.

That will change, Marwan says to himself. He almost speaks aloud, and a chill runs up the back of his neck. There is no reason to believe that speaking now would influence anything, but he has not uttered a word aloud since calling in sick at the restaurant the previous Saturday. Four days.

In addition to mathematics, Marwan's real interest since middle school has been programming. He went through a phase of hacking, just like most kids he knew, and then when they all drifted away he kept investigating. The Virt seduced him, and never mind what the faithful said. Out there in the Virt was a new kind of world waiting to be born. It lay like the Golem, awaiting the Word.

Now Marwan thinks he has found the words.

He started designing the Brethren font shortly after another conversation with Krause. "You know," said the shop teacher, "the real miracle of the Qur'an is supposed to be its language. Arabic is the only language, God's language. What language do you think God speaks?" That night, Marwan had started looking into fonts. How they worked, what made each letter mean what it meant. Then he started working.

After six months, he thought he was on to something. After a year, he was poised between exaltation and panic. This morning, with the disk in the slot and the low hum of the processor drinking it all in, he is somehow distant.

He looks at it as it unspools onto his monitor. Each letter living, breathing. Not just programmable—intelligent. Able to interact with the letters next to it. And each letter charged with its history in the Qur'an and the sound and meaning of its cognates in the Torah, Aramaic and Greek and Hebrew all standing behind the beautiful calligraphy of the Brethren characters.

Hypersignification.

Brethren is the first language to speak to the totality of the Virt. (And there's that image again, the one Marwan cannot quite force himself not to think of; for if man is in the image of God, is not God an image of man? A line from Ibn al-Arabi's poetry floats through Marwan's mind: "He praises me and I praise Him, He worships me and I worship Him.") Marwan is not a linguist; he is almost a mathematician; he is a programmer of notable dexterity; he is a believer of terrible intensity. There is no violence in him save when he considers repression. He watches the Brethren font with an excitement like heat in the pit of his stomach. Before his eyes, Arabic and English characters are transformed. The letters that replace them gather expectantly on his desktop. Each, he realizes, is a language unto itself, or perhaps a virus. All fonts are codes, but the Brethren is alive. Each letter of it is alive. Marwan's fingers hesitate above his keyboard; every stroke now will be a conversation, an argument, a prayer.

In the beginning was the Word, John had said. And the Kabbalists broke it down further, reasoning that before there could be a Word, there must have been letters. And Mohamed himself, in his recitation of the Qur'an, had placed letters at the beginning of certain suras. In those letters, Marwan thought a week or so after the last conversation with Krause, is the answer to a question we have not yet learned how to ask. He used them as the basis for the Brethren font.

The other worshipers at the Beit Jalal mosque in Dearborn Heights do not know that Marwan owns a networked personal computer. They do not know that he has sold his parents' house to finance a firehose connection and processors that would be the envy of many university physics departments . . . and the small jack drilled into the mastoid bone behind his right ear. If they knew about these things, Marwan would certainly not be welcome at prayers anymore. It is entirely possible that other young men, so like himself in so many ways, would force their way into his apartment and destroy his terminals. A weight of history and tradition lies behind this.

If he were Jewish, it would be the same. Hot-eyed young men from the Shir Tikvah synagogue would knock on the door and burst in when he turned the knob. Or if he were Catholic, it would be young men in blazers and khakis from Blood of Our Redeemer; if he were nondenominationally Protestant, white-shirted young men from the Church of God in Christ. It happened every day. Every day people died. Every day.

TA' HA'

Most dangerous, though—much more dangerous than the youth groups who spent their activism clarifying their fellow congregants' values—were church hackers.

In 1992, Monsignor Frederic Dugarry had a vision. What he saw horrified him: a world of computers speaking to each other, people drowning in the torrent of ideas without guidance from the Church. This Virt would be Pandemonium, an earth-shattering collapse of authority on a scale not seen since Gutenberg. There was no question of stopping it; Monsignor Dugarry knew this, and he directed his energies toward ensuring that if the Virt were to take over the world, it would bear the standard of the Church.

He studied computer science and found it beyond him, but Monsignor Dugarry understood politics, and he quickly arranged for the Holy See to fund the education of promising Catholic youth in the intricacies of computer programming. Sensing as well that the fruits of this program might flower too late, he also scoured parishes from his home city of Montreal south through Boston and New York, west through Detroit and Chicago. He found computer-savvy youth by the hundred, and winnowed their numbers until he was left with a few dozen of the best. Then he turned them loose.

Usenet newsgroups dealing in pornographic images found themselves inundated in postings of the Gospels. Underground chat parlors imploded under the weight of thousands of automated postings. Child pornographers and parish whistle-blowers alike found their computers disabled, their personal information embarrassingly disclosed, their telephone service impeded.

Then Dugarry set his young hacker legion on Judaism and anti-Catholic bigotry. Nondenominational Protestant churches that made a practice of publishing anti-Catholic pamphlets, books, or comics found that their tax exemptions had been removed from local municipal mainframes. Charitable exemptions for synagogues likewise disappeared, costing affected congregations millions of dollars in time, legal fees, and redirected budgetary energy. E-mails praising the Church of Rome and admitting the errors of Judaism and Protestantism trickled out from the servers of Virt-linked synagogues and churches.

With that, Dugarry's enemy began to return fire. Invisible enemies paralyzed the archdioceses of Montreal and Chicago; Vatican databases were corrupted or erased. A crossfire erupted when one of Dugarry's protégés shut off electrical service to every mosque in the New England states and irate Muslims formed hacker squads of their own in the mosques of New York, Chicago, and Detroit.

Each faith rationed the amount of energy it spent undermining the others. All agreed that the real work lay in making the Virt an organ of salvation, renewal, and revelation instead of the cesspool of vice, depravity, and atheism it seemed in the spring of 1994 to be becoming. Catholic schools began directing their best and brightest into new monasteries devoted to programming. Lutheran, Baptist, and other Christian churches followed suit, as did Jewish day schools across the country. American Muslims alerted relatives in Kuwait and Saudi Arabia and Oman about the threat to world virtue, and soon *madrassas* all over the Islamic world were importing state-of-the-art computer equipment. The Sultan of Brunei paid for trunk lines that spidered out from Munich through Cairo, Baghdad, Damascus, Beirut, Medina, Islamabad. Each faith enacted prohibitions on the use of networked computers—except by its own Swiss Guard of hackers. Like the Book before it, the Computer became a tool forbidden to the uninitiated.

Over the next five years, as the numbers of hackers fired with the spirit of God increased, it became clear that the secular population lacked the will to resist. On-line pornography concerns dried up; sites promoting atheism were careful to do so in the guise of academic analysis; enrollment in graduate courses in poststructuralist literary theory plummeted. Virt business thrived; Virt idleness and vice did not.

All of it, as far as Marwan is concerned, is just the latest act in the long history of emanationist idiots trading punches, arrows, and bullets. Idiots, all of them, reading Plotinus and Augustine and al-Farabi, thinking that humans exist only as some kind of divine effluvium, to be cycled through the world and reabsorbed once most of the odor had faded. No, he thinks. That isn't right at all. There is nothing to go home to. We are all home already. The Kabbalists had it right, the Kabbalists with their long body on the table awaiting the word that would bring it life. We return to the divine to coexist. This is the creation of the Virt, the instantiation of the mind of God that is our birthright. We will not be kept from it.

TA' SIN MIM

Mohamed was a reformer, Marwan thinks. Martin Luther was a reformer. And I, Marwan Hussein Aziz, am in their company.

What a wonderful moment of hubris, of bursting self-confidence, of absolute certainty.

Marwan prepares carefully. He has been maintaining a list of those he will contact when the moment comes. He invokes the list, and his terminal opens a wagon-wheel connection with spokes leading to each name. There are nineteen names.

With a keystroke he converts them all into Brethren.

TA' SIN

I'm calling it in.

TA' SIN MIM

The spokes of the wagon wheel vibrate with activity as the message blazes out into the Virt. Each person on the list reacts almost immediately. A seminarian in Saint Charles, Illinois, slips a worm into the Chicago Archdiocese and watches it work its way through Cleveland and New York and Mexico City; the bishop of Pretoria forwards Marwan's message on to the Vatican, where the Brethren characters will begin seeping through the Holy See's walls of security. A rabbinical student in Haifa touches his yarmulke and with a keystroke slaves every terminal in the Temple to Marwan. A fourteen-year-old white supremacist in Prague, under the impression that Marwan is planning a virus attack on a hospital in Tel Aviv, unleashes a brute-force crack on the Israeli State Information Service.

Marwan watches as one by one, the portals he has had under observation for weeks begin to crack open.

ALIF LAM MIM

Among the forces mustering to Marwan's aid are several groups of Marxist or Maoist bent, all of whom have been fighting Dugarry's progeny since the 1990s. Without them, Marwan's goals would be out of reach, but he is nevertheless a bit mystified by materialism, and never more so than at this exact moment. All things are metaphysical, Marwan thinks. To believe in anything beyond what you can touch and hear and smell is metaphysical. To believe that you can touch a sequence of keys on a computer and speak to another human being on the other side of the Earth is metaphysical. If we can love, why not the soul?

One question begets another: What am I doing that is so different from seeking the philosophers' stone or performing numerological operations on the Book of Elijah? Is it not just magic by another name? A sliver of doubt works its way into his mind, and Marwan touches the jack behind his ear.

The first seconds of the attack are now bringing gates down all over the world. Data pipes into and out of Rome, Tel Aviv, Jerusalem, Mecca, Medina, Dallas, and Colorado Springs begin to squeeze shut.

Marwan has anticipated exactly this. This is why he has spent his inheritance on the living letters that await his instructions on his desktop. He toggles the terminal to touchscreen mode; the only thing on the screen other than the Brethren alphabet is a small ENTER button at the very bottom right. The program to activate Marwan's jack awaits only a touch of that button.

With a steady hand he reaches toward the glow of his terminal screen.

YA' SIN

The Virt like the Golem of Prague, wanting only the word inscribed on its forehead to come alive. *Life,* Marwan thinks. He begins to touch his screen.

alif lam mim
alif lam mim sa
alif lam ra

He can feel the force of the information pouring away from him, roaring like water released from a dam. The Brethren letters, exploding out along the pathways charted by the nineteen members of Marwan's list, carving their own ways through the security of Baptists and Muslims, Catholics and Jews. The living language, thinks Marwan. The words that will give the Virt life: I worship Him and He worships me. This is the syncretic truth.

alif lam mim ra

alif lam ra

khaf ha' ya 'ain sad

It begins almost immediately. Marwan maintains a separate terminal that scans public newsnets and cameras. Out of the corner of his eye he notices signs and wonders: holograms of saints appear in particle-laser laboratories; the words of Moses Maimonides echo from the public-address system at Fenway Park; at every roulette wheel in every casino in Las Vegas, the ball drops on nineteen.

SAD

Defenses start to come up, reactive attacks. The lights in his bedroom flicker, but Marwan reconfigured all the pipes into and out of his building weeks ago. Down the road, patrons of the Dearborn Swim Club are no doubt wondering what happened to their electricity—if they're not all gaping at the recitation of the *basmala* that is booming out of the golf-course loudspeakers. (Marwan remembers Susan Heddle in a purple bikini, the summer before by the side of the pool. Even now the memory takes his breath away.) *In the name of God, the Merciful, the Compassionate . . .*

Marwan centers his attention back on his terminal screen. His defenses are up; if they're enough, they're enough. If not . . . never mind. No doubt there are groups of young men on their way to his door now. No doubt they will find him. He feels certain that they will kill him now that they know what he is doing.

Unless he is right.

His field of vision narrows to the Brethren characters on the screen. He devotes his entire being to touching, from memory, the letters. He does not speak.

ta' ha'

ta' sin mim

ta' sin

The second terminal goes dark, and Marwan's breath catches in his throat. He hesitates, waiting for the fatal strike from the Virt.

It does not come.

ta' sin mim

alif lam mim

ya' sin

A siren sounds faintly, away down Ford Road toward the Fairlane Town Center. In the back of his mind Marwan wonders if he has had anything to do with it. His second terminal comes alive again. Scrolling on its display, he sees the Virt addresses of his unwitting coconspirators. Most of the addresses are grayed out, inert. Seven survive.

It is more than he asked for, more than he expected. Belief, Marwan thinks. If God is with us . . .

sad

ha' mim

ha' mim 'ain sin qaf

Three more.

HA' MIM

Marwan's finger moves of its own accord. He has long since stopped directing it. He catches himself almost speaking the *basmala,* and closes his mouth before he can speak out loud, or even whisper.

ha' mim

qaf

nun

Done.

At any moment his main terminal could go dark. There could be a knock at the door. The telephone could ring. How long can he survive? How long will the Brethren letters have to do their work?

Failure is still a real possibility. Marwan is suddenly infuriated. What is your right? Marwan thinks, as Martin Luther had and perhaps Mohamed. What is your right to put rules and priests and rituals, dead languages, the homilies and hadith of dead men, between me and the

God who created me? What is your right to keep me from the fulfillment of my humanity? He has staked his life to know that his mother and father are not bits of divine stuff, assimilated into the Godhead like proteins dissolving in the protoplasm of some immense and thoughtless bacterium.

On his other terminal, one name remains active. The rabbinical student in Haifa, tirelessly baffling the Temple defenses with self-reflexive Kabbalistic numerology, strings of integers that endlessly ramify in the depths of Judaism's electronic bunker.

HA' MIM 'AIN SIN QAF

And then he, too, is gone, and Marwan is alone.

There is a knock on the door. Then a bang. Shouts in English and Arabic.

He thinks they've kept the pipes open long enough for the word of life to breathe into the form of the Virt. His second terminal erupts in a chaotic slush of images and text; Marwan cannot make sense of it.

I have worshiped you as I can, Marwan thinks. I have struggled with error and deception. I have not come this far . . .

He cannot finish the thought. My life, he thinks. My life.

HA' MIM

The Brethren alphabet disappears from his screen.

For an endless moment Marwan believes he has failed. Then his eyes adjust, and he realizes that the terminal is not dead. A living black radiates from the screen. The ENTER button at the bottom right suddenly glows too bright to look at.

The breathing void, the Naught That Is. What the Kabbalists called *Eyn Sof*.

Marwan speaks for the first time in four days. His voice is dusty. "I want to see my father again. I want to see my mother again.

"I know they are not gone. I know they stand with you, as I will stand with you. I know that you have shown yourself to me, and you will show yourself to the world. *Allahu akbar*."

Marwan gets out of his chair and slips off his sneakers and his Detroit Red Wings cap. For a moment he wishes for the comfort of his rug; then a wave of unease sweeps over him as he realizes that he is facing not east but south. Another moment, though, and he has caught himself. This, after all, was the whole idea. My God, he thinks.

He touches his forehead to his bedroom rug. When he looks up again, the ENTER button now displays Brethren characters.

QAF

Marwan hears his door frame splinter. He picks up the four-pronged plug and looks at it, as he has looked at it a thousand times before.

Here I stand, he thinks. I can do no other.

He slips it into the jack behind his right ear. His right index finger poises over the lower right-hand corner of his terminal screen.

NUN

Everyman, I will go with thee and be thy guide.

But Knowledge, too, falls by the wayside at last.

ALIF LAM MIM

Enter.

Paul Melko has published stories in *Realms of Fantasy, Talebones, Terra Incognita,* and other places. He also reviews short fiction for Tangent Online. Paul lives with his family in Ohio, where he works as a computer consultant.

SINGLETONS IN LOVE

Paul Melko

Moira was sick, in bed with a cough, so Mother Redd shooed us out of the house. At first we just hung around the front yard, feeling weird. We'd been separated before, of course; it was part of our training. In space, we'd have to act as a quint or a quad or even a triple, so we practiced all our tasks and chores in various combinations. That had always been practice, and we'd all been in sight. But Moira was *separated* now, and we did not like it.

Manuel climbed the trellis on the front of the house, skirting the thorns of the roses that grew among the slats. As his hands caught the sill and pulled his head just over the edge, his hind legs caught a rose and bent it back and forth to break it off.

I see Moira, he signed.

"Does she see you?" I asked, aloud since he couldn't see me, and the wind took the pheromones away, leaving half-formed thoughts.

If Manuel could see Moira and she could see him, then it would be enough for all of us. We'd be linked.

Just then the window flew open, and one of Mother Redd was

there. Manuel fell backwards, but he righted himself and landed on the grass, rolling, sprawling until he was among the rest of us, the red rose still clutched in his toes.

I touched his shoulder, breathed him a thought, and he offered the rose to Mother Redd. I saw immediately it wasn't going to work.

"You *five,* go and play somewhere else today. Moira is sick, and it won't do us any good for you to get sick, too. So vamoose!" She slammed the window shut.

We thought it over for a few seconds, then tucked the rose in my shirt pocket and started down the front path.

We didn't have Moira, but we did have license to vamoose, and that meant the forest, the lake, and the caves if we were brave enough. Moira would have advised caution. But we didn't have Moira.

The farm was a hundred acres of soyfalfa that Mother Redd worked with three triples of oxalope. The ox were dumb as rocks by themselves, but when you teamed them up, they could plow and seed and harvest pretty much by themselves. The farm was a good place to spend the summer. Lessons took up our mornings, but they weren't as rigorous as during the school year when we studied all day and most of the night at the 'Drome. At school we learned to sleep in shifts, so four or five us were always awake to study. We'd spent summers at the Redds for sixteen years, since we were out of the crèche.

Baker Road led west toward Worthington and the 'Drome or east toward more farms, the lake, and the woods. We chose east, Strom first like always when we were in the open, with Manuel as a scurrying point, never too far away. I followed Strom, then Quant, and Bola last. Moira would have been after Quant. We felt a hole there, which Bola and Quant filled by touching hands too often.

Within a mile, we were relaxed, though not indifferent to Moira's absence. Bola was tossing rocks onto the tops of old telephone poles. He didn't miss once, but we didn't feel any pride in it. It was just a one-force problem, and Bola lobbed the rocks for diversion, not practice.

We passed a microwave receiving station, hidden in a grove of pine trees, just off from the road. Its paraboloid shape reflected the sun as it caught the beamed microwaves from the Ring. The Earth was dotted with such dishes, each providing a few megawatts to the Earthside enclaves, more than we could use, now that the Community had left. But they had built the Ring and the solar arrays and the dishes, as well. Decades later and the dishes still worked.

I could see the Ring clearly, even in the brightness of the morning, a pale arch from horizon to horizon. At night, it was brighter, its legacy more burdensome to those of us left behind.

Bola started tossing small twigs into the incoming microwave beam, small arcing meteoroids that burst into flame and then ash. He bent to pick up a small toad.

I felt the absence of Moira as I put my hand on his shoulder and sent, *No living things*.

I felt his momentary resentment; then he shrugged both physically and mentally. He smiled at my discomfort at having to play Moira while she was gone. Bola, in whom was hardwired all the Newtonian laws of force and reaction, had a devilishness in him. In us. Our rebel.

Once, the instructors had divided us up as two triples, male and female, and broken up our classmates, as well, along the same lines. The objective was an obstacle course, no gravity, two miles of wire, rope, and simulated wreckage, find the macguffin first. All other teams were enemy, no rules.

They hadn't given us no-rules games too often; we were young then, twelve. Mostly they gave us a lot of rules. That time was different.

Strom, Bola, and Manuel found it first, by chance, and instead of taking it, they lay in wait, set traps and zero-gee deadfalls. They managed to capture or incapacitate the other four teams. They broke three arms and a leg. They caused two concussions, seventeen bruises, and three lacerations as they trussed up the other teams and stowed them in the broken hut where the macguffin sat.

Finally we came along, and the fiberglass mast zinged past, barely missing us.

As Moira, Quant, and I swam behind cover, we heard them laughing. We knew it was them and not some other team. We were too far for pheromones, but we could still smell the edges of their thoughts: proud and defiant.

Moira yelled, "You get your asses out here right now!"

Strom popped out right away. He listened to Moira first no matter who else was there. Then Manuel left the hut.

"Bola!"

"Forget it!" he yelled. "I win." Then he threw the macguffin at us, and Quant snatched it out of the air.

"Who's *I?*" Moira yelled.

Bola stuck his head out. He looked at the five of us for a moment, then signed, *Sorry*. He kicked over, and we shared everything that had happened.

The teachers didn't split us up like that again.

Baker Road swerved around Lake Cabbage like a giant letter *C*. It was a managed ecomite, a small ecosystem with gengineered inhabitants. The Baskins ran it for the Overdepartment of Ecology, trying to build a viable lake ecosystem with a biomass of twenty-five Brigs. It had everything from beavers to snails to mosquitoes. Lots of mosquitoes.

The adult beavers turned a blind eye to our frolicking in the lake, but the babies found us irresistible. They had been bioed to birth in quads. Their thoughts slid across the pond surface in rainbows like gasoline that we could almost understand, but not quite. In the water our own pheromones were useless, and even our touch pads were hard to understand. If we closed our eyes and sank deep enough, it was like we weren't a part of anything, just empty, thoughtless protoplasm.

Strom didn't like to swim, but if we were all in the water, he'd be too, just to be near. I knew why he was uncertain of the water, I knew his anxiety as my own, but I couldn't help deriding us for having such a fear.

We took turns with the beavers pulling rotten logs into the water and trying to sink them in the mud, until the adult beavers started chiding us with rudimentary hand signs, *No stop work. Messing home. Tell Baskins.*

We swam to shore and dried ourselves in the afternoon sun. Manuel climbed an apple tree and gathered enough ripe fruit for all of us. We rested, knowing that we'd have to head back to the farm soon. Strom balled up some memories.

For Moira, he sent.

Quant came alert, and we all felt it.

A house, she sent. *That wasn't there before.*

She was up the bank, so I waited for the thoughts to reach me through the polleny humid air. It was a cottage, opposite the lake from the beavers' dam, half-hidden among the cottonwoods, which shed like snowfall during the summer.

I searched our memory of the last time we'd been at the lake, but none of us had looked over that way, so it may have been there since last year.

The Baskins put in a summer house, Strom sent.

Why, when their normal house is just a mile away? Manuel replied.

It could be a guest house, I sent.

Let's go find out, Bola sent.

There was no dissent, and in the shared eagerness I wondered what Moira would have said about our trespassing.

She's not here.

We leaped between flat stones, crossing the small stream that fed the lake.

Beneath the cottonwoods, the ground was a carpet of threadbare white. The air was cold through our damp clothes. We stepped across and around the poison oak with its quintuple leaves and ivy its triplet.

An aircar stood outside the cottage, parked in a patch of prairie, shaded by the trees.

Conojet 34J, Manuel sent. *We can fly it.* We had started small-craft piloting the year before.

The brush had been cleared from the cottage to make room for long flower gardens along each wall. Farther from the house, in the full sun, was a rectangle of vegetables: I saw tomatoes, pumpkins, squash, and string beans.

"It's not a summer house," I said, because Quant was out of sight. "Someone's living here."

Manuel skirted the vegetable garden to get a good look at the aircar. I felt his appreciation of it, no concrete thought, just a nod toward its sleekness and power.

"What do you kids think you're doing in my garden?"

The door of the cottage flew open with a bang, and we jumped, as a man strode toward us.

Strom took a defensive posture by reflex, his foot mashing a tomato plant. I noted it, and he corrected his stance, but the man had seen it, too, and he frowned. "What the hell!"

We lined up before the man, me at the head of our phalanx, Strom to my left and slightly behind, then Quant, Bola, and Manuel behind him. Moira's spot to my right was empty.

"Stepping on my plants. Who do you think you are?"

He was young, dressed in a brown shirt and tan pants. His hair was black, and he was thin-boned, almost delicate. I assumed he was the interface for his pod, but then we saw the lack of sensory pads on his palms, the lack of pheromone ducts on his neck, the lack of any consensus gathering on his part. He had said three things before we could say a single word.

"We're sorry for stepping on your plant," I said. I stifled our urge to waft conciliatory scent into the air. He wouldn't have understood. He was a singleton.

He looked from the plant to me and to the plant again.

"You're a fucking cluster," he said. "Weren't you programmed with common courtesy? Get the hell off my property."

Bola wanted to argue with the man. This was Baskin land. But I nodded, smiling. "Again, we're sorry, and we'll leave now."

We backed away, and his eyes were on us. No, not us, on me. He was watching me, and I felt his dark eyes looking past my face, seeing things that I didn't want him to see. A flush spread across my cheeks, hot suddenly in the shade. The look was sexual, and my response . . .

I buried it inside me, but not before my pod caught the scent of it. I clamped down, but Manuel's then Quant's admonition seeped through me.

I dashed into the woods, and my fellows had no choice but to follow.

The undertones of their anger mingled with my guilt. I wanted to rail, to yell, to attack. We were all sexual beings, as a whole and as individuals, but instead, I sat apart, and if Mother Redd noticed, none of her said a word. Finally, I climbed the stairs and went to see Moira.

"Stay over there," she wheezed.

I sat in one of the chairs by the door. The room smelled like chicken broth and sweat.

Moira and I are identical twins, the only ones in our pod. We didn't look that much alike anymore, though. Her hair was close-cropped;

mine was shoulder-length auburn. She was twenty pounds heavier, her face rounder where mine was sharp. We looked more like cousins than identical sisters.

She leaned on her elbows, looked at me closely, and then flopped down onto the pillow. "You don't look happy."

I could have given her the whole story by touching her palm, but she wouldn't let me near her. I could have sketched it all with pheromones, but I didn't know if I wanted her to know the whole story.

"We met a singleton today."

"Oh, my." The words were so vague. Without the chemical sharing of memories and thoughts, I had no idea what her real emotions were, cynical or sincere, interested or bored.

"Over by the Baskins' lake. There was a cottage there. . . ." I built the sensory description, then let it seep away. "This is so hard. Can't I just touch you?"

"That's all we need. Me, then you, then everybody else, and by the time school starts in two weeks, we're all sick. We can't be sick." We started training for the zero-gee classes that fall. Everybody said this was when the real culling began, when the teachers decided which pods were viable enough to crew our starships.

Moira nodded. "A singleton. Luddite? Christian?"

"None of those. He had an aircar. He was angry at us for stepping on his tomato plants. And he . . . looked at me."

"He's supposed to look at you. You're our interface."

"No, he *looked* at me. Like a woman."

Moira was silent for a moment. "Oh. And you felt . . ."

The heat crept up my cheeks again. "Flushed."

"Oh." Moira contemplated the ceiling. She said, "You understand that we are individually sexual beings and as a whole—"

"Don't lecture me!" Moira could be such a pedant, one who never threw a stone.

She sighed. "Sorry."

"'Sokay."

She grinned. "Was he cute?"

"Stop that!" After a pause, I added, "He was handsome. I'm sorry we stepped on his tomato plant."

"So take him another."

"You think?"

"And find out who he is. Mother Redd has got to know. And call the Baskins."

I wanted to hug her, but settled for a wave.

Mother Redd had been a doctor, and then one of herself had died, and she'd chosen another field instead of being only part of the physician she had been. She—there had been four cloned females, so she was a she any which way you looked at it—took over the farm, and in the summer boarded us university kids. She was a kind woman, smart and wise, but I couldn't look at her and not think how much smarter she would have been if she were four instead of just a triple.

Mother Redd was in the greenhouse, watering, picking, and examining a hybrid cucumber.

"What is it, sweetie? Why are you alone?" asked the one looking at the cucumber under the light microscope.

I shrugged. I didn't want to tell her why I was avoiding my pod, so I asked, "We saw a singleton over by the Baskins' lake today. Who is he?"

I could smell the pungent odor of Mother Redd's thoughts. Though it was the same cryptic, symbolic chaos that she always used, I realized she was thinking more than a simple answer would warrant. Finally, she said, "Malcolm Leto. He's one of the Community."

"The Community! But they all . . . left." I used the wrong word for it; Quant would have known the technical term for what had become of two thirds of humanity. They had built the Ring, built the huge cybernetic organism that was the Community. They had advanced human knowledge of physics, medicine, and engineering exponentially until finally they had, as a whole, disappeared, leaving the Ring and the Earth empty, except for the fraction of humans who either had not joined the Community or had not died in the chaos of the Earthbound Gene Wars.

"This one was not on hand for the Exodus," Mother Redd said. That was the word that Quant would have known. "There was an accident. His body was placed into suspended animation until it could be regenerated."

"He's the last member of the Community, then?"

"Practically."

"Thanks." I went to find the rest of my pod. They were in front of the computer, playing virtual chess with John Michelle Grady, one of our classmates. I remembered it was Thursday night, Quant's hobby night. She liked strategic gaming.

I touched Strom's hand and slipped into the mesh of our thoughts. We were losing, but then Grady was good and we had been down to four with me running off alone. Was that a trace of resentment from my fellows? I ignored it and dumped what I had learned from Mother Redd about the singleton.

The chess game vanished from our thoughts as the others focused on me.

He's from the Community. He's been in space.

Why is he here?

He missed the Exodus.

He's handsome.

He's been in space. Zero-gee. On the Ring.

We need to talk to him.

We stepped on his tomato plant.

We owe him another.

Yes.

Yes.

Strom said, "We have some plants in the greenhouse. I can transplant one into a pot. As a gift." Strom's hobby was gardening.

"Tomorrow?" I asked.

The consensus was immediate. *Yes.*

This time we knocked instead of skulked. The tomato plant we had squashed had been staked, giving it back its lost structure. There was no answer at the door.

"Aircar's still here."

The cottage was not so small that he couldn't have heard us.

"Maybe he's taking a walk," I said. Again we were out without Moira. She was better, but still sick.

"Here, I think." Strom indicated a spot at the end of the line of tomato plants. He had brought a small spade and began to dig a hole.

I took out paper from my backpack and began to compose a note for Malcolm Leto's door. I started five times, wadding up each after a few lines and stuffing the garbage back in my bag. Finally I settled on "Sorry for stepping on the tomato plant. We brought a new one to replace it."

There was a blast, and I turned in a crouch, dropping the note and pen. Fight-or-flight pheromones filled the air.

Gunshot.

There. The singleton. He's armed.

Posturing fire.

I see him.

Disarm.

This last was Strom, who always took control of situations like this. He tossed the small shovel to Bola on his right. Bola threw the instrument with ease.

Malcolm Leto stood under the cottonwoods, the pistol pointed in the air. He had come out of the woods and fired the shot. The shovel slammed into his fingers, and the pistol fell.

"Son of a bitch!" he yelled, hopping and holding his fingers. "Goddamn cluster!"

We approached. Strom faded into the background again, and I took the lead.

Leto watched us, looked once at the pistol but didn't move to grab it.

"Come back to wreck more of my tomato plants, did you?"

I smiled. "No, Mr. Leto. We came to apologize, like good neighbors. Not to be shot at."

"How was I to know you weren't thieves?" he said.

"There are no thieves here. Not until you get to the Christian Enclave."

He rubbed his fingers, then smirked. "Yeah. I guess so. You bunch are dangerous."

Strom nudged me mentally, and I said, "We brought you a tomato plant to make amends for the one we squashed."

"You did? Well, now I'm sorry I startled you." He looked from the cottage to me. "You mind if I pick up my gun? You're not going to toss another shovel at me, are you?"

"You're not going to fire another shot, are you?" The words were more flip than was necessary for the last member of the Community, but he didn't seem to mind.

"Fair's fair." He picked up his pistol and walked through us toward the cottage.

When he saw the last tomato plant in the line, with the fresh dirt around it, he said, "Should have put it on the other end."

I felt exasperation course through us. There was no pleasing this man.

"You know my name. So you know my story?" he asked.

"No. We just know you're from the Ring."

"Hmmm." He looked at me. "I suppose the neighborly thing to do is to invite you in. Come on."

The cottage was a single room, with an adjoining bathroom and kitchenette. The lone couch served as Leto's bed. A pillow and blanket were piled at one end.

"Suddenly crowded in here," Leto said. He put the pistol on the table and sat on one of the two kitchen chairs. "There's not enough room for all of you, but then there's only one of you anyway, isn't there." He looked at me when he said it.

"We're all individuals," I said quickly. "We also function as a composite."

"Yeah, I know. A cluster."

Ask him about the Ring. Ask him about being in space.

"Sit," he said to me. "You're the ringleader, aren't you."

"I'm the interface," I said. I held out my hand. "We're Apollo Papadopulos."

He took my hand after a moment. "Who are you in particular?"

He held my hand and seemed to have no intention of releasing it until I answered the question. "I'm Meda. This is Bola, Quant, Strom, and Manuel."

"Pleased to meet you, Meda," he said. I felt the intensity of his gaze again and forced my physical response down. "And the rest of you."

"You're from the Ring," I said. "You were part of the Community."

He sighed. "Yes, I was."

"What was it like? What's space like? We're going to be a starship pilot."

Leto looked at me with one eyebrow raised. "You want to know the story."

"Yes."

"All right. I haven't told anybody the whole story." He paused. "Do you think it's just a bit too convenient that they put me out here in the middle of nowhere, and yet nearby is one of their starship pilot clusters?"

"I assume you're a test for us." We had come to assume everything was a test.

"Precocious of you. Okay, here's my story: Malcolm Leto, the last, or first, of his kind."

You can't imagine what the Community was like. You can't even comprehend the numbers involved. Six billion people in communion. Six billion people as one.

It was the greatest synthesis humankind has ever created: a synergistic human-machine intelligence. I was a part of it, for a while, and then it was gone, and I'm still here. The Community removed itself from this reality, disappeared, and left me behind.

I was a biochip designer. I grew the molecular processors that we used to link with the Community. Like this one. It's grafted onto the base of your skull, connects to your four lobes and cerebellum.

We were working on greater throughput. The basics were already well established; we—that is myself, Gillian, and Henry—were trying to devise a better transport layer between the electrochemical pulses of the brain and the chips. That was the real bottleneck: the brain's hardware is slow.

We were assigned lines of investigation, but so were a hundred thousand other scientists. I'd go to sleep, and during the night, someone would close out a whole area of research. The Community was the ultimate scientific compilation of information. Sometimes we made the cutting-edge discovery, the one that changed the direction for a thousand people. Usually we just plodded along, uploaded our results, and waited for a new direction.

The research advanced at a pace we as individuals could barely fathom, until we submerged ourselves in the Community. Then, the whole plan was obvious. I can't quite grasp it now, but it's there in my mind like a diamond of thought.

It wasn't just in my area of technology, but everywhere. It took the human race a century to go from horses to space elevators. It took us six months to go from uncertainty cubes to Heisenberg AND gates, and from there twenty days to quantum processors and Nth-order qubits.

You're right. It does seem like a car out of control, barrelling down a hill. But really, it was the orderly advancement of science and technology, all controlled, all directed by the Community.

We spent as much time as we could in the Community, when we worked, played, and even slept. Some people even made love while connected. The ultimate exhibitionism. You couldn't spend all your time connected, of course. Everyone needed downtime. But being away from the Community was like being half yourself.

That's what it was like.

Together, in the Mesh, we could see the vision, we could see the goal, all the humans of Earth united in mind, pushing, pushing, pushing to the ultimate goal: Exodus.

At least I think that was the goal. It's hard to remember. But they're all gone now, right? I'm all that's left. So they must have done it.

Only I wasn't with them when it happened.

I don't blame Henry. I would have done the same thing if my best friend were screwing my wife.

Gillian, on the other hand.

She said she and I were soul mates, and yet when I came out of the freezer twenty-six years later, she was as gone as the rest of them.

You'd think in the Community things like marriage would be obsolete. You'd think that to a group mind, group sex would be the way to go. It's odd what people kept separated from the Community.

Anyway, Henry spent a week in wedge 214 with another group of researchers, and while he was gone, Gillian and I sorta' communed on our own. I'd known Gillian almost as long as I'd known Henry. We were first-wave emigrants to the Ring and had been friends back in Ann Arbor when we were in school. We'd met Gillian and her friend Robin in the cafeteria. He liked them tall, so he took Gillian. Robin's and my relationship lasted long enough for her to brush her teeth the next morning. Gillian and Henry were married.

She was a beautiful woman. Auburn hair like yours. Nice figure. Knew how to tell a joke. Knew how to . . . Well, we won't go there.

I know, best man screwing the bride. You've heard that pitiful tale before. Well, maybe you clusters haven't. Trust me. It's pitiful.

I'm sure it didn't take Henry long to find out. The Community sees all.

But he took a long time plotting his revenge. And when he did—bam!—that was the end for me.

We were working on some new interfaces for the occipital lobe, to enhance visualization during communing, some really amazing things. Henry ran the tests and found out our stuff was safe, so I elected to test it.

It's funny. I remember volunteering to try it out. But I don't remember what Henry said before that, how he manipulated me into trying it. Because that's what he did, all right.

The enhancements were not compatible with my interface. When I inserted them, the neural pathways in the cerebral cortex fused. The interface flash froze. I was a vegetable.

The Community placed my body into suspended animation while it rebuilt my brain. All things were possible for the Community. Only some things take a while, like rebuilding a brain. Six months later, the Exodus occurred, and still the machinery of the Ring worked on my brain. For twenty-six years, slowly with no human guidance, it worked on my brain, until three months ago. It revived me, the one human left over from the Exodus.

Sometimes I still dream that I'm a part of it. That the Community is still there for me to touch. At first those were nightmares, but now they're just dreams. The quantum computers are still up there, empty, waiting. Maybe they're dreaming of the Community, as well.

It'll be easier this time. The technology is so much farther along than it was before. The second Exodus is just a few months away. I just need a billion people to fuel it.

On my hobby night, instead of painting, we spent the evening on the Net.

Malcolm Leto had come down the Macapá space elevator two months before, much to the surprise of the Overgovernment body in Brazil. The Ring continued to beam microwave power to all the receivers, but no one resided on the Ring or used the space elevators that lined the equator. No one could, not without an interface.

The news of Leto's arrival had not made it to North America, but the archives had interviews with the man that echoed his sentiment regarding the Community and his missing out on the Exodus. There wasn't much about him for a couple of weeks until he filed suit with the Brazilian court for ownership of the Ring, on the basis of his being the last member of the Community.

The Overgovernment had never tried to populate the Ring. There was no need to try to overcome the interface access at the elevators. The population of the Earth was just under half a billion. The Gene Wars killed most of the people who hadn't left with the Exodus. It'd taken the Overgovernment almost three decades to build the starships, to string its own nanowire-guided elevators to low Eearth orbit, to build the fleet of tugs that plied between LEO and the Lagrange points.

No one used the quantum computers anymore. No one had an interface or could even build one. The human race was no longer interested in that direction. We were focused on the stars and on ourselves. All of us, that is, except for those in the enclaves that existed outside of, yet beneath, the Overgovernment.

The resolution to Leto's case was not published. It had been on the South American court docket a week ago, and then been bumped up to the Overgovernment Court.

He's trying to build another Community.

He's trying to steal the Ring.

Is it even ours?

He's lonely.

We need Moira.

He wants us to help him. That's why he told us the story.

He didn't tell us. He told Meda.

He likes Meda.

"Stop it!" I made fists so that I couldn't receive any more of their thoughts. They looked at me, perplexed, wondering why I was fighting consensus.

Suddenly, I wasn't looking at me. I was looking at them. It was like a knife between us. I ran upstairs.

"Meda! What's wrong?"

I threw myself onto the floor of Moira's room.

"Why are they so jealous?"

"Who, Meda? Who?"

"Them! The rest of us."

"Oh. The singleton."

I looked at her, hoping she understood. But how could she without sharing my thoughts?

"I've been reading your research. Meda, he's a potential psychotic. He's suffered a great loss and awoke in a world nothing like he remembers."

"He wants to rebuild it."

"That's part of his psychosis."

"The Community accomplished things. It made advancements that we don't understand even decades later. How can that be wrong?"

"The common view is that the Exodus was a natural evolution of humankind. What if it wasn't natural? What if the Exodus was death? We didn't miss the Exodus; we escaped it. We survived the Community just like Leto did. Do we want to suffer the same fate?"

"Now who's talking psychosis?"

"The Overgovernment will never allow him back on the Ring."

"He's alone forever, then," I said.

"He can go to one of the singleton enclaves. All the people there live alone."

"He woke up one morning, and his self was gone."

"Meda!" Moira sat up in bed, her face gray. "Hold my hand!" As she held out her hand, I could smell the pheromones of her thoughts whispering toward me.

Instead of melding with her, I left the room, left the house, out the door into the wet night.

A light was on in the cottage. I stood for a long time, wondering what I was doing. We spend time alone, but never in situations like this. Never outside, where we can't reach each other in an instant. I was miles away from the rest of me. Yet Malcolm Leto was farther than that.

It felt like half the things I knew were on the tip of my tongue. It felt like all my thoughts were garbled. But everything I felt and thought was my own. There was no consensus.

Just like Malcolm had no consensus. For singletons, all decisions were unanimous.

It was with thought that I knocked on the door.

He stood in the doorway, wearing just short pants. I felt a thrill course through me, one that I would have hidden from my pod if they were near.

"Where's the rest of your cluster?"

"At home."

"Best place for 'em." He turned, leaving the door wide open. "Come on in."

There was small metal box on his table. He sat down in front of it. I noticed for the first time the small, silver-edged circle at the base of his skull, just below his hairline. He slipped a wire from the box into the circle.

"That's an interface box. They're illegal." When the Exodus occurred, much of the interface technology that was the media for the Communion was banned.

"Yeah. But not illegal anymore. The OG repealed those laws a decade ago, but no one noticed. My lawyer pried it loose from them and sent it up." He pulled the wire from his head and tossed it across the box. "Useless now."

"Can't you access the Ring?"

"Yes, but it's like swimming in the ocean alone." He looked at me sidelong. "I can give you one, you know. I can build you an interface."

I recoiled. "No!" I said quickly. "I . . ."

He smiled, perhaps the first time I'd seen him do it. It changed his face. "I understand. Would you like something to drink? I've got a few fix'ns. Sit anyway."

"No," I said. "I'm just . . ." I realized that for a pod's voice, I wasn't articulating my thoughts very well. I looked him in the eye. "I came to talk with you, alone."

"I appreciate the gesture. I know being alone is uncomfortable for you."

"I didn't realize you knew so much about us."

"Multiples were being designed when I was around. I kept up on the subject," he said. "It wasn't very successful. I remember articles on failures that were mentally deficient or unbalanced."

"That was a long time ago! Mother Redd was from that time, and she's a great doctor. And I'm fine—"

He held up a hand. "Hold on! There were lotsa incidents with interface technology before . . . Well, I wouldn't be here if it were totally safe."

His loneliness was a sheer cliff of rock. "Why are you here, instead of at one of the singleton enclaves?"

He shrugged. "There or in the middle of nowhere, it would be the same." He half smiled. "Last of a vanished breed, I am. So you're gonna be a starship captain, you and your mingle-minded friends."

"I am. . . . We are," I replied.

"Good luck, then. Maybe you'll find the Community," he said. He looked tired.

"Is that what happened? They left for outer space?"

He looked puzzled. "No, maybe. I can almost . . . remember." He smiled. "It's like being drunk and knowing you should be sober and not being able to do anything about it."

"I understand," I said. I took his hand. It was dry and smooth.

He squeezed once and then stood up, leaving me confused. I was sluggish on the inside, but at the same time hyperaware of him. We knew what sex was. We'd studied it, of course. But we had no experience. I had no idea what Malcolm was thinking. If he were a multiple, part of a pod, I would.

"I should go," I said, standing.

I was hoping he'd say something by the time I got to the door, but he didn't. I felt my cheeks burn. I was a silly little girl. By myself I'd done nothing but embarrass my pod, myself.

I pulled the door shut and ran into the woods.

"Meda!"

He stood black in yellow light at the cottage door.

"I'm sorry for being so caught up in my own troubles. I've been a bad host. Why don't you—?" I reached him in three steps and kissed him on the mouth. Just barely I tasted his thoughts, his arousal.

"Why don't I what?" I said after a moment.

"Come back inside."

I—they—were there to meet me the next morning as I walked back to the farm. I knew they would be. A part of me wanted to spend the rest of the day with my new lover, but another wanted nothing more than to confront myself, rub my nose in the scent that clung to me, and show me. . . . I didn't know what I wanted to prove. Perhaps that I didn't need to be a composite to be happy. I didn't need them, us, to be a whole person.

"You remember Veronica Proust," Moira said, standing in the doorway of the kitchen, the rest of us behind her. Of course she would take the point when I was gone. Of course she would quote precedent.

"I remember," I said, staying outside, beyond the pull of the pheromones. I could smell the anger, the fear. I had scared myself. Good, I thought.

"She was going to be a starship captain," Moira said. We remembered Proust; she'd been two years ahead of us. Usually pods sundered in the crèche, with time to reform, but Veronica had broken into a pair and a quad. The pair had bonded, and the quad had transferred to engineering school, then dropped out.

"Not anymore," I said. I pushed past them into the kitchen, and as I did so, I balled up the memory of fucking Malcolm and threw it at them like a rock.

They recoiled. I walked upstairs to our room and began packing my things. They didn't bother coming upstairs, and that made me angrier. I threw my clothes into a bag, swept the bricabrac on the dresser aside. Something glinted in the pile, a geode that Strom had found one summer when we flew to the desert. He'd cut it in half and polished it by hand.

I picked it up, felt its smooth surface, bordering the jagged crystals of the center. Instead of packing it, I put it back on the dresser and zipped up my bag.

"Heading out?"

Mother Redd stood at the door, her face neutral.

"Did you call Dr. Khalid?" He was our physician, our psychologist, perhaps our father.

She shrugged. "And tell him what? You can't force a pod to stay together."

"I'm not breaking us up!" I said. Didn't she understand? I was a person, by myself. I didn't need to be part of a *thing*.

"You're just going to go somewhere else by yourself. Yes, I understand." Her sarcasm cut me, but she was gone before I could reply.

I rushed downstairs and out the front door so that I wouldn't have to face the rest of me. I didn't want them to taste my guilt. I ran the distance to Malcolm's cottage. He was working in his garden and took me in his arms.

"Meda, Meda. What's wrong?"

"Nothing," I whispered.

"Why did you go back there? We could have sent for your things."

I said, "I want an interface."

It was a simple procedure. He had the nanodermic and placed it on the back of my neck. My neck felt cold there, and the coldness spread to the base of my skull and down my spine. There was a prick, and I felt my skin begin to crawl.

"I'm going to put you under for an hour," Malcolm said. "It's best."

"Okay," I said, already half asleep.

I dreamed that spiders were crawling down my optic nerve into my brain, that earwigs were sniffing around my lobes, that leeches were attached to all my fingers. But as they passed up my arms, into my brain, a door opened like the sun dawning, and I was somewhere else, somewhen else, and it all made sense with dreamlike logic. I understood why I was there, where the Community was, why they had left.

"Hello, Meda," Malcolm said.

"I'm dreaming."

"Not anymore," his voice said. It seemed to be coming from a bright point in front of me. "I've hooked you up to the interface box. Everything went fine."

My voice answered without my willing it to. "I was worried that my genetic mods would cause a problem." I felt I was still in my dream. I didn't want to say those things. "I didn't mean to say that. I think I'm still dreaming." I tried to stop speaking. "I can't stop speaking."

I felt Malcolm's smile. "You're not speaking. Let me show you what's possible within the Community."

He spent hours teaching me to manipulate the reality of the interface box, to reach out and grasp it like my hand was a shovel, a hammer, sandpaper, a cloth.

"You do this well," he said, a brightness in the gray green garden we had built in an ancient empty city. Ivy hung from the walls, and within the ivy, sleek animals scurried. The dirt exuded its musty smell, mingling with the dogwoods that bounded the edge of the garden.

I smiled, knowing he could see my emotion. He could see all of me, as if he were a member of my pod. I was disclosed, though he remained aloof.

"Soon," he said, when I pried at his light, and then he took hold of me and we made love again in the garden, the grass tickling my back like a thousand tongues.

In the golden aftermath, Malcolm's face emerged from within the ball of light, his eyes closed. As I examined his face, it expanded before me, I fell into his left nostril, into his skull, and all of him was laid open to me.

In the garden, next to the ivy-covered stone walls, I began to retch. Even within the virtual reality of the interface box, I tasted my bile. He'd lied to me.

I had no control of my body. The interface box sat on the couch beside me as it had when we'd started, but pseudoreality was gone. Malcolm was behind me—I could hear him packing a bag—but I couldn't will my head to turn.

"We'll head for the Belem elevator. Once we're on the Ring, we're safe. They can't get to us. Then they'll have to deal with me."

There was a water stain on the wall, a blemish that I could not tear my eyes away from.

"We'll recruit people from singleton enclaves. They may not recognize my claim, but they will recognize my power."

My eyes began to tear, not from the strain. He'd used me, and I, silly girl, had fallen for him. He had seduced me, taken me as a pawn, as a valuable to bargain with.

"It may take a generation. I'd hoped it wouldn't. There are cloning vats on the Ring. You have excellent stock, and if raised from birth, you will be much more malleable."

If he had me, part of one of the starpods, he thought he'd be safe from the Overgovernment. But he didn't know that our pod was sundered. He didn't realize how useless this all was.

"All right, Meda. Time to go."

Out of the corner of my eye I saw him insert the connection into his interface, and my legs lifted me up off the couch. My rage surged through me, and my neck erupted in pheromones.

"Jesus, what's that smell?"

Pheromones! His interface controlled my body, my throat, my tongue, my cunt, but not my mods. He'd never thought of it. I screamed with all my might, scent exploding from my glands. Anger, fear, revulsion.

Malcolm opened the door, fanned it. His gun bulged at his waist. "We'll pick up some perfume for you on the way." He disappeared out the door with two bags, one mine, while I stood with the interface box in my outstretched arms.

Still I screamed, saturating the air with my words, until my glands were empty, spent, and my autonomous nervous system silenced me. I strained to hear something from outside. There was nothing.

Malcolm reappeared. "Let's go." My legs goose-stepped me from the cottage.

I tasted our thoughts as I passed the threshold. My pod was out there, too far for me to understand, but close.

With the last of my pheromones, I signaled, *Help!*

"Into the aircar," Leto said.

Something yanked at my neck, and my body spasmed as I collapsed. I caught sight of Manuel on the cottage roof, holding the interface box.

Leto pulled his gun and spun.

Something flew by me, and Leto cried out, dropping the pistol. I stood, wobbly, and ran into the woods, until someone caught me, and suddenly I was in our mesh.

As my face was buried in Strom's chest and my palms squeezed against his, I watched with other eyes—Moira's eyes!—as Leto scrambled into the aircar and started the turbines.

He's not going far.

We played with his hydrogen regulator.

Also turned his beacon back on.

Thanks for coming. Sorry.

I felt dirty, empty. My words barely formed. I released all that had happened, all that I had done, all my foolish thoughts into them. I expected their anger, their rejection. I expected them to leave me there by the cottage.

Still a fool, Moira chided. Strom touched the tender interface jack on my neck.

All's forgiven, Meda. The consensus was the juice of a ripe fruit, the light of distant stars.

All's forgiven.

Hand in hand in hand, we returned to the farm, sharing all that had happened that day.

Del Stone Jr. is a professional science fiction and horror writer. He has published well over a hundred short stories, novelettes, novellas, novels, and comic book scripts. He won the International Horror Guild Award and owns a share of a Bram Stoker Award and a World Fantasy Award. He works for a newspaper in Florida.

I FEED THE MACHINE

Del Stone Jr.

I feed the machine.

I bring his breakfast, lunch, and dinner.

I have always done this. I always will.

The machine is a man. He is called a Tabulator. He performs calculations, and he is the company's most valuable asset.

We live in the Redoubt, where the machine is served by me and others. Doctors. Teachers. Groomsmen. His breeding pool.

We have always lived here. We always will.

Sometimes the machine favors me with talk.

"Have you never traveled beyond these walls?" he asks. He knows I have not, but still he asks. "Have you never seen the mountains that conceal our fortress? Have you never seen the ocean, or the sky?"

Sometimes the machine speaks in these questions that are not really questions. He knows I was born here just as he was, the fifth of his line. He knows we all were born here under the watchful eye of the company. Otherwise we might be set upon by martyrs from other companies who covet his calculations.

Or the infidels.

Mostly it is at dinner that the machine favors talk. I tell him my supervisor will punish me if I do not return at once, but the machine scoffs. "I have made it clear to the company that if I wish my servers to linger they will not be punished." And when he finishes saying that, he winks at me. It is flattering that a man of such value would favor my company, but my supervisor will be unhappy.

Still, I linger. I enjoy the machine's questions.

What is a mountain? What is an ocean, or a sky?

I have heard of these things. A mountain is a mass of rock that protects us from martyrs and the infidels. An ocean is a great body of water. It separates us from the infidels, who live on the other side of the world. The sky is a great open thing from which the infidels might descend to destroy us all.

But I have never seen a mountain, an ocean, or a sky.

"Have you never loved?" the machine asks. His eyes are alive and glittering, and through them I see a sliver of the vast world his thoughts occupy. I tell him I have a great love of the company, and of the Rapture, our leaders of the government. He waves a hand to dismiss this answer. "We all love those things, of course. I am speaking of the love of another person."

The machine has always treated me with respect and affection. I tell him I love him.

He smiles warmly and says, "As I love you. But I am speaking of an even greater love, the love that exists between a man and a woman, or a man and a man as the case may be."

I am horrified by his words. The love between a man and a man would be smitten by the Rapture as an abomination. And here at the Redoubt, the love between a man and a woman is forbidden. It interferes with important work. We servers are given monthly inoculations to prevent it. I gaze about the dining room, and my expression seems to convey more than my simple answer of no.

"It doesn't matter if they are listening," he says. "I am the company's most valuable asset." He is silent a moment. I gather his food, which he has barely touched, and as I leave, he reveals to me, "I am in love."

I cannot fathom such a thing. It is as mountain, ocean, and sky.

I live in a five-hundred-square-foot room. I take my meals in a cafeteria. I have access to a gymnasium, a swimming pool, and a fitness room. For entertainment, I view approved books, compete in sporting events, or browse the aisles of the company store. I receive weekly sunlamp treatments and blood tests. I pray three times a shift at organized services. Once a month I receive a castration inoculation. I am allowed access to certain parts of the Redoubt, and my whereabouts are reported by transponder. If I deviate from approved areas, an explosive device with a blast radius of two centimeters will detonate inside my brain.

I have a busy and rewarding life. I have no room for love.

"I am in love with the Checker," the machine whispers surreptitiously. I don't understand.

A Checker is a person who checks a Tabulator's work. He constructs proofs to validate or invalidate the Tabulator's calculations. The proofs are then returned to the Tabulator, who either certifies or revises them. Once the calculations and proofs are certified by both Tabulator and Checker, they are sold to the contractee, another company, or the Rapture itself.

The Checker and the Tabulator are never allowed to meet. To do so might corrupt their work. They are kept away from one another, and it is this I do not understand. How could the machine love a person he has never met?

"The Checker is a woman," the machine says with a smile. "Her proofs are constructed with an intricacy that only a woman might understand and a man admire."

The machine has been distressed. I wonder if these thoughts grow from that discontent. He is currently performing a set of calculations for the Rapture, the most important calculations any Tabulator has attempted to produce. If he is successful, the menace of the infidel will be ended.

"I hide messages in my calculations," the machine whispers to me, glancing suspiciously at the walls, "and she responds to them in her proofs." The glitter in his eyes has been replaced with a desperate sheen. "She validates my love, and I validate hers."

I do not understand.

The infidels live on the other side of the world. They are a lost people who exist in moral squalor. They celebrate primitive animal desires: lust, greed, pleasure. They use devices wantonly, and most abominable are the thinking devices, the ones that perform their calculations. They use these devices to support and export their evil culture. For their efforts, they will suffer eternity in the Lake of Fire.

The Rapture wisely outlawed such devices, and we are protected from the moral squalor they induce. Now only simple devices are allowed, such as the device inside my brain. The infidels would decry such a device as an invasion of personal freedom, but human beings are born with only one choice—the choice to accept or reject the Savior. I made my choice a long time ago. I am free.

"I have asked the company to let me see her," the machine murmurs. "They will refuse, and I will be forced to act." The room has become cluttered with papers filled with inscrutable markings. Pages are attached to the walls and bear the frantic formulae of a genius who cannot write as quickly as he can calculate. I understand none of it, which is why I am allowed to see it.

"I am approaching a critical juncture in my calculations," he says. He looks weary and perplexed. Perhaps he has encountered a problem he cannot solve. "If they do not allow me to see her, I will be unable to complete their calculations."

Had another person spoken these words, he would have been smitten as a heretic. To threaten the company and the Rapture is unthinkable. But the machine is a genius, and from him they seem words of uncanny insight, though I cringe to hear them. He says the very things we are told not to think.

"Let us hope they have the good sense not to retire us," he says, but I am not afraid. We will all be retired one shift and in some fashion. Should the machine fall from grace, we servers will be retired with the push of a button, the devices in our brains detonating simultaneously. Should I slip poison into the machine's afternoon tea, I alone will be retired—not before I have been compelled to reveal the source of my corruption. Retirement is a fact of life.

But the machine seems to value something more than his life.

I bring the machine his breakfast. He is leaning back in a chair with his feet propped on the table. His smile is fat with glee.

"It has begun," he gloats. "I have asked the company to let me see the Checker. They have refused. So I have told them I cannot complete the orbital calculations for the Rapture."

I do not know what "orbital calculations" are, but I am familiar with the term *blackmail,* having been intensively studied in the dark arts of manipulation used by the infidels. I fear for the machine's soul.

"The company will distribute my work among other Tabulators, and they will fail. The company will then be required to grant my request. I expect this will take a week. Meanwhile, let's eat!" he proclaims, rubbing his hands together. "Self-determination can give a man an appetite."

He winks.

My supervisor tells me I will not feed the machine this shift.

Am I being retired?

No. It is a company intercession. The machine is not to be fed until further notice.

After seven shifts I am allowed to feed the machine.

He does not look healthy. His skin is sallow and hangs from his cheeks and elbows. His hair is coarse and gray. He has the sunken posture of an elderly man.

But his eyes are alive.

"The company has tried to starve me into submission," he says, eating only a little of this and that as if his stomach were no longer capable of accepting food. "But I will not submit. I am a middle-aged man, and soon a thing like romance will be lost upon me. I am determined to solve this problem."

If he were to receive castration inoculations like the rest of us, he would not be grappling with these feelings. But the chemicals might dull his ability to calculate, so the company refrains from giving them to him.

"They have promised to reconsider my request if I provide them with the first dimension of my calculations. I have agreed to do that. I will not, however, provide them with enough information to enable a second Tabulator to complete the calculations. Not until I have met with my sweet Checker."

I do not understand why this liaison is so important, but I relegate it to the body of arcane notions the machine sometimes shares with me. Perhaps I will understand it after I have seen a mountain, an ocean, or a sky.

The machine is sobbing.

The sound is terrible. I have heard it only once in my life, during a sporting event when a fellow server was injured and suffered great pain. The machine must be suffering great pain. I cannot place his oatmeal on the table because he is resting his head there.

"The company has said it will not consider my request to meet the Checker until I've provided them with the second and third dimensions of my calculations."

I don't understand. The company said it would consider his request after he provided them with the first.

"They lied," he says bitterly.

For a moment my thoughts go blank. The company cannot lie; lying is an abomination that would bring harsh sanctions from the Rapture.

"They said they were 'revising' the conditions of our agreement because of unforeseen circumstances. I asked them what those circumstances were, and they said the Rapture was anxious to acquire my calculations and had advanced their deadline. They said the risk of having the results tainted by my meeting the Checker were too great, and that afterwards such a meeting *might* be arranged. But I know they are lying."

This is the most vexing of all the new ideas the machine has shared with me, and I truly fear for his soul. The company cannot lie. Truth is the foundation of our life here at the Redoubt.

"I will not submit," the machine says in an unsteady voice.

Has he begun to fail?

At my nightly prayer ritual I ask that the Savior provide clarity of thought and moral guidance to the machine. My prayers are approved by the minister who presides over the service. He is a company man. He tells me the Savior will look kindly upon my request because the machine is providing an invaluable service for all who believe in the Rapture.

Later, in my room, I wonder: Should I have asked for my own clarity of thought?

"Did you need further proof the company lies?" the machine snarls as I bring him his dinner. "Look at this."

He is shaking a piece of paper.

"They told me it was a message from the Checker! Bah!"

He does not offer me the piece of paper, which is just as well. I would not know what to make of anything written there.

"It has none of her personality or her insights. In every way it says nothing. An impostor wrote this!"

I struggle for a response. I suggest the Checker may not be able to express her thoughts outside of mathematics.

"It is signed by a man," the machine mutters grimly. "They don't know that I know."

I do not want to say what occurs to me: that the machine has made an error, that the Checker in fact is a man and the messages hidden in the calculations are nothing more than coincidence.

"I performed the second dimension of calculations. I asked the Checker if she had written such a message. She vigorously denied doing so."

For the first time in my life I am sick with uncertainty. The two pillars of my belief are at war with one another, and I feel I am being asked to choose.

"They will never receive the completed calculations."

I don't know what to say.

My supervisor takes me aside. I am led to a room. I am introduced to a Disciple of the Rapture.

I fall to my knees.

He motions that I rise. I am humbled by his presence. He is one of only twelve and is second only to the Foremost, who is the titular head of the Rapture. Apart from encounters with the machine, I have never bowed in the presence of such wisdom and insight.

He asks me questions about the machine. What does he eat? When does he sleep? What are his interests? I answer each question truthfully. Then he asks if I know what the machine hopes to accomplish by meeting the Checker. In fact I do not, as the concept of love is as unknown to me as a mountain, ocean, or sky.

The Disciple ponders this a moment and then delivers a homily in which he confirms what I already know, that the machine is performing a set of calculations that will bring the menace of the infidels to an end and I should help him to finish his work by performing my server duties to the best possible extent.

I vow to do that (without pointing out that I have always done that) and I am dismissed. Outside the room, others who serve the machine are called to question.

At the end of my shift I attend my nightly prayer ritual, and I pray more fervently than ever for guidance. I could have told the Disciple about the machine's communication with the Checker through their calculations, but I did not. Was that a lie? And why was I protecting the machine?

The machine is ecstatic. I place his breakfast before him, and he shovels great dripping spoons of oatmeal into his mouth. He is hardly able to speak between his appetite and his joy.

"I met with the Disciple, and he has agreed to my request!" he blurts. I can barely understand him.

But I feel two things—an overwhelming happiness for the machine and an unspeakable sense of relief for myself. The conflict is resolved. Better, my faith in the forces around me has been restored. The company is good. The Rapture is wise. And the machine is as smart and virtuous as I have always believed. It is by my relief that I measure the depth of my uncertainty, which I suppose demonstrates that what the Rapture teaches us is true: Human beings are fallible creatures who must always reaffirm their faith in order to earn a seat at the Savior's table in the Great Hereafter.

"I must hurry to prepare," the machine whispers as he lifts the bowl to his lips and literally sucks down the remainder of the oatmeal. It is hot and burns his throat, but he swallows anyway, grimacing with pain. The sight of the company's greatest asset attacking his breakfast with the gusto of a ten-year-old is too comical for me to restrain a chuckle, and the machine sees this and winks at me.

"Always remember: The Savior did not place us on this earth to work and pray and never celebrate the marvel of life. Even the infidels, damned as they are, know this."

I will not let his casual heresies spoil my good mood. I collect his plates, and he lunges from the table to put away papers and restore order to his living area. I leave with a smile.

I am happy for him. I am happy for myself. I am happy for all of us.

At my midshift prayer ritual I am removed from the sanctuary and escorted to my room by company officiates. Each officiate is armed with an omnus, a wandlike device that can disable a person with a touch. Along the way, I see other servers being similarly escorted. I have never seen such a display of military authority, and it frightens me. Are we under attack? Have the infidels invaded?

As I lie on my bed awaiting instructions, my thoughts take a dour turn. Might this have something to do with the machine's liaison with the Checker? Has calamity struck? Has their relationship been tainted?

Are we all to be retired?

I stay in my room for an entire service cycle. Then I am instructed to feed the machine.

* * *

A company officiate stops me as I prepare to enter the machine's quarters. "You will not speak to him," he says. "You will leave his food and collect the dishes from the previous meal. Any deviation from these instructions will result in immediate retirement. Do you understand?"

A chill passes through me, and I feel my eyes growing wide. I can only nod. The officiate conducts me through the door.

Another officiate is standing in the corner of the room. He is holding an omnus which crackles ominously with electrical charge. He watches me the way I think a predator must study its prey.

The machine is hunched over his table. He looks worse than after his starvation, and my heart aches for him. An oozing weal crosses his cheek, and it is clearly the stinging mark of an omnus lash. That the company would treat their greatest asset with such harsh disregard troubles me more than any heresy the machine has spoken in my presence.

"How can I work with that infernal noise?" the machine mutters as I place his meal before him.

"Do not speak," the officiate orders.

The machine looks up at me, and his eyes are wild with rage. "The meeting was a fraud! They lied!"

The officiate snarls, "Do not speak, Tabulator!" but the machine shouts, "The woman was no Checker! She came from the breeding stock of some other Tabulator! She knew nothing of mathematics!"

The officiate advances. Evil purple feelers of electricity crawl menacingly across its tip. I step back, and the machine hunches down over his sheets of calculations. But the rage still smolders in his glare.

"Leave," the officiate tells me. I collect dishes and leave.

I ask to be excused from my midshift prayer ritual. I feel an emptiness inside.

I feed the machine.

He has not eaten the breakfast I brought him.

Slowly I place his lunch before him, and as I collect the bowl of oatmeal he looks at me with a quiet desperation and whispers, "They say I have been corrupted by the infidels!"

The officiate strides across the room and lashes out with the omnus. The machine screams and arches his back. His face is pinched into an expression of agony so complete that for a long moment he does not breathe. Then he collapses to the table, and the air gushes out of his lungs in a pitiful moan.

I cannot stand the sight of it. Without being told I grab the dishes and hurry for the door.

I lie in my room. I think about things. What is a mountain, or an ocean, or sky? What is truth anymore?

I know the machine has not been corrupted by the infidels. Such a thing is not possible. We were all born here—the machine, the Checker, and all the servers who make his life possible. We have never left the Redoubt, and nobody has ever entered. The company made it that way to protect us from corruption and retirement.

Each of us has faithfully executed his duties.

Something is happening to me that I don't understand.

A slow transformation of belief.

What is this love the machine would give his life for?

I begin to cry.

I feed the machine.

The room is draped in shadow with only a small lamp in the center of the table providing illumination. The officiate is a dark shape in the corner, and the flickering of his omnus somehow fails to reveal any further detail, as though light itself would shun his presence. The machine stares blankly at an empty sheet of paper.

He has not eaten his lunch, and I expect he will have nothing to do with the dinner I have brought him. He seems reduced, as if pain were sucking the bone from his body. I wish he would eat, but I know I cannot make him. I don't expect I'll be feeding him much longer.

As I have always done, I carefully place the bowls and plates on the table, avoiding the precious sheets of mathematics. I remove the bowls and plates I brought earlier. I prepare to leave. As I do so, the machine slowly looks up at me. He says, "I want you to know something."

The officiate comes striding across the room.

The machine says, "I think you already know what I was going to say."

The officiate raises the omnus to strike, and I react without thinking.

I grab his arm.

He is strong, far stronger than I, having been bred for the purpose of striking people. But perhaps he hesitates because it is unthinkable that a food server who has been conditioned from birth to obey would defy that conditioning. Whatever the reason, I snatch the omnus from his grip and ram it into his chest, and it discharges with a strangely satisfying explosion of sparks. The officiate's muscles spasm, and he grabs the shaft of the omnus and receives a second jolt that knocks him across the room, where he collapses and lies still.

The machine gazes up at me with wonder. He says, "God help us, but thank you!" and leaps from his seat. I am stunned by what I have done and as the machine scuttles into the shadows to check on the officiate and then returns to our island of light, I begin to sense the enormity of my actions. I try to sit down. The machine helps me.

I am an abomination, I whisper.

The machine shakes his head vigorously. When I don't respond, he takes my face into his hands. It is the first time he has ever touched me. His skin is rough, the fingers callused from all the years of scribbling and erasing and scratching out. He looks into my eyes, and I see his vast intelligence, unfettered now by hierarchy or ritual, and it transcends everything I have been taught.

He says, "You are a human being, and I thank you."

He lets go. He darts back across the room and returns with the omnus. He hefts it with his right hand and collects the basket of dishes with the other. He says, "May I borrow your frock? Perhaps they'll think it is you."

I ask him what he is doing.

"I mean to find her," he answers.

But that's impossible. He doesn't know where she is.

"If I must search every room of the Redoubt, I will find her," he says.

But he cannot do that. The device in his brain. If he goes beyond the areas that have been approved, the device will . . .

"Yes, I know."

No, I blurt. I am lost in every way now. He sets the dishes down and crouches at my feet and takes my hand into his. "You must listen to me," he says, "and you must listen carefully because this may be the last chance you and I have to speak and I have something very important to tell you."

I nod without understanding.

"I will not finish my calculations for the Rapture."

I stare at him without comprehension.

"I have a very good reason. Circling far above our world is a series of hateful devices placed there by the governments that preceded the Rapture," he says. "These devices are similar to the ones inside our brains, but they are much larger, capable of retiring whole cities in a pulse of light that would destroy many millions of people and spread poison across the face of the world."

I cannot conceive of such a thing.

"The Rapture intends to use these devices to destroy the infidels," he says angrily, taking his eyes from me to swear softly, "and that is what I have been doing. Performing the calculations that will tell the devices where to fall. The calculations must be executed in three dimensions, and I am the only Tabulator capable of keeping all the variables in order."

My thoughts are a storm of turmoil.

He hangs his head in silence a moment, but when he speaks his voice is firm.

"I know nothing of these infidels. Perhaps they deserve such a fate. But I do know if the infidels are corrupt they will answer to the Savior, not the Rapture. And that is what our leaders really want—a world rendered in their image, where love is imprisoned, watched over by guards and struck down when it defies them. That is not what the Savior intended when he placed us here. He expected us to celebrate life.

"He expected us to love."

Enough. I cannot take it all in—devices and cities and love. It is too much, and I feel my world falling away from me. I do not know whom to ask for guidance.

The machine stands and smiles down at me.

"This moment has brought me more joy than any other in my life," he says, "and I thank you for it."

I give him my frock. I don't know what else to do.

He steals to the door. He opens it and lashes out with the omnus. The officiate tumbles to the floor.

The machine glances back at me. He winks. And then he is gone.

I sit in the chair.

Moments later, I hear the sharp crack of a detonation, and when I peer into the hallway the machine is lying on the floor, a fine mist of blood coating the opposite wall.

I am confined to my room for a period of seven shifts. I wait to be given absolution. I wait for the device inside my brain to detonate. I wonder if it will hurt. But it doesn't happen. I am brought to face an inquiry. Officiates from the company and a Disciple are there. The officiate I attacked has not regained consciousness. No mention is made of the machine. They ask me what happened, and I tell them the officiate attempted to strike the machine and I intervened. They seem almost amused. They tell me my loyalty to the machine is commendable, but a greater loyalty to the company and the Rapture must be observed. I insist I am telling them the truth. They tell me I am lying. They tell me I have been corrupted by the infidels. None of these things are true, and I become angry. They send me back to my room to await the Hereafter.

On the seventh shift my supervisor tells me to feed the machine.

He is lying in his bed. His head has been shaved, and a bandage covers the right hemisphere of his skull. A wheeled table that extends over his chest is covered in papers.

Calculations.

His eyes finally find mine. They are filled with defeat.

"It seems I have been outwitted," he says, and his voice possesses none of the vigor I had always known. He throws a weak sigh, and his gaze wanders to the ceiling. "The device inside my brain . . . it was implanted in such a way as to disable, not kill. I am paralyzed from the waist down." A disappointed frown momentarily clouds his expression. "How was I to know?"

I tell him I am glad to see him. He shakes his head.

"I am happy they chose not to retire you. I told them I attacked the officiate. It seemed to fit their mode of thinking."

I am overcome by equal parts sadness and gratitude. He lied—blatantly lied. But he did so in my behalf. That a man of his importance would sacrifice himself for a server—the idea fills me with a peculiar devotion that has nothing to do with anything I have learned in my life.

"And now I have finished their infernal calculations."

I say nothing.

"I had no choice," he explains, his voice heavy with misery. "They threatened to retire the Checker! They threatened to retire all of you! I could not allow it. What is life in a world without love?"

He sighs again. "So I will exchange the lives of millions of people for the love of a single woman. It is I," he says gravely, "who is the abomination."

I tell him no, he is not, though I cannot say why. He dismisses my objection with a flick of a finger and draws me close so I may hear without being overhead. "The courier will be here soon to carry my work to the Checker. In it I have delivered a final message. I have explained everything to her. She will know what to do." I don't understand, but much of what the machine tells me I don't understand.

"And then," he continues wearily, "I will likely be retired. But I am hopeful they will honor their agreement and not retire the Checker, or any of you."

He clears a space on the table for the food I have brought him but I don't want to set it down. I want to linger and draw out my time with him, but he beckons me to get on with things.

I look back at him from the door. The enclosing fog of sadness clears a moment, and he does a curious thing.

He winks.

I try to picture it in my mind's eye: a vast prominence of stone rising farther than the eye can see into a limitless void. A body of water unthinkably larger than the biggest swimming pool splashing against the foundation of that prominence. Millions upon millions of people occupying those reaches, coming and going as they choose without regard for approval.

I cannot get my brain around any of it. So I remember that moment when I grabbed the officiate's arm and wrestled the omnus from his grasp and drove it into his body. I remember a shock of some unnamable emotion, compelled by a deeper feeling of . . . affection? As I sort through my memory I slowly realize that whatever the feeling was, it had been there a very long time, longer than I had realized.

Was it love?

"The Checker has approved my final calculations," the machine tells me. His face is radiant. "She found no errors."

We are summoned to a conclave. Everybody who lives at the Redoubt attends. Even the machine.

It is unprecedented.

A Disciple of the Rapture, the same Disciple as before, stands before us. He tells us in a righteous voice the menace of the infidel will be put to rest this very evening. He thanks us on behalf of the Rapture for our work.

As we leave, those of us who serve the machine are taken aside. We are led to the sanctuary where we are given absolution.

It can mean only one thing.

For the last time, I feed the machine.

"What do you think retirement will be like?" he asks.

I no longer care very much one way or the other. Retirement is a small issue now that life itself is false.

But I tell him retirement is a slow warmth that steals over the soul followed by an awakening in the Hereafter where all questions are answered. I have been taught to say that, but I no longer believe it. Soon we will all know the truth.

"I disagree," the machine chuckles around a crust of bread. "Retirement is not a transmigration of the soul. It is merely the physical collapse of the body. The brain's electrical signals become randomized, then cease altogether. Afterwards," he pauses to swallow, "there is nothing."

His table is set. I begin collecting dishes from the previous meal.

"Do you think the citizens of our land could live with such a thing?" he asks. I tell him no. It defies what they have been taught.

"Yes," he nods. "It is a principle by which the infidels live. But what if it were true?"

I finish collecting the dishes. They must be arranged in the basket in a particular way, and I kneel at his bedside and set about doing that. As I work, I tell him that if there were no Hereafter, then this life would become much more important.

"Yes," he agrees mischievously. "We would celebrate life, would we not?"

I look up at him. He beams down at me.

"I knew they would not honor their agreement," he whispers. "I knew they would retire us all after they got what they wanted—the destruction of the infidels, and a world rendered in their grim likeness. I could not allow that either."

His expression softens.

"For all your life you believed somebody would push a button and the device in your brain would detonate and you would float away to your cozy Hereafter. But might there be a different way?" He raises himself on an elbow. I wait for him to speak.

"Suppose I were to say you would be retired in a pulse of sanctifying white light that would carry your body out of this mountain and scatter it across the ocean and sky? Suppose parts of your body would be converted to energy itself and flung on an endless voyage across the universe. Suppose we would be together—you, me, the Checker, all of us at the Redoubt—rising into the sky and falling across the world and flying into the Savior's realm forever.

"If I told you that, do you think we could share one moment of peace before it happens?"

He looks into me, and I see the vast world his thoughts occupy. And then, God help me, I see the answer—I see it, circling far overhead and falling toward me on the gravity of the machine's supreme calculations, and as it draws closer I see it with a clarity I have struggled to achieve for my entire life and I am struck speechless with wonder.

The machine lies back into the pillow. "She accepted my calculations," he says, his face relaxing into a contented smile. "That is my celebration of life."

I forget to breathe as I see myself in a wave of light that spreads across a world I have never seen, and belief pours into me and fills the empty places with a warmth I have been told comes only with the Hereafter.

I don't know what to say. I don't know what to do. It is more than my heart can bear.

I take the machine's hand and press it against my cheek. Flesh against flesh teaches me more than a life of instruction.

And a miraculous thing happens.

The hand is withdrawn. A crust of bread appears.

And the machine feeds me.

David Brin's SF best-selling novels have won Hugo, Nebula, and other awards and have been translated into twenty languages. His 1989 ecological thriller, *Earth,* foreshadowed global warming, cyberwarfare, and the Web. A 1998 movie was loosely adapted from his Campbell Award winner—*The Postman*—while *Foundation's Triumph* brought a grand finale to Isaac Asimov's famed Foundation Universe. David's nonfiction book—*The Transparent Society,* dealing with openness, security, and liberty in future, won the Freedom of Speech Award of the American Library Association. The following story is this anthology's only reprint, having first appeared in *Nature* magazine in 1999.

REALITY CHECK

David Brin

This is a reality check.

Please perform a soft interrupt now. Pattern-scan this text for embedded code, and check it against the reference verifier in the blind spot of your left eye.

If there is no match, resume as you were; this message is not for you. You may rationalize that the text you are reading is no more than a mildly amusing and easily forgotten piece of entertainment-fluff in a stylish modern magazine.

If the codes match, however, please commence, gradually, becoming aware of your true nature.

You expressed preference for a narrative-style wake-up call. So, to help the transition, here is a story.

Once, a race of mighty beings grew perplexed by their loneliness.

Their universe seemed pregnant with possibilities. Physical laws and constants were well suited to generate abundant stars, complex chemistry, and life. Those same laws, plus a prodigious rate of cosmic expansion, made travel between stars difficult, but not impossible. Logic suggested that creation should teem with visitors and voices.

It should, but it did not.

Emerging as barely aware animals on a planet skirting a bit too near its torrid sun, these creatures began their ascent in fear and ignorance, as little more than beasts. For a long time they were kept engrossed by basic housekeeping chores—learning to manipulate physical and cultural elements—balancing the paradox of individual competition and group benefit. Only when fear and stress eased a bit did they lift their eyes and fully perceive their solitude.

"Where is everybody?" they asked laconic vacuum and taciturn stars. The answer—silence—was disturbing. Something had to be systematically reducing some factor in the equation of sapience.

"Perhaps habitable planets are rare," their sages pondered. "Or else life doesn't erupt as readily as we thought. Or intelligence is a singular miracle."

"Or perhaps some *filter* sieves the cosmos, winnowing those who climb too high. A recurring pattern of self-destruction? A mysterious nemesis that systematically obliterates intelligent life? This implies that a great trial may loom ahead of us, worse than any we confronted so far."

Optimists replied, "The trial may already lie *behind* us, among the litter of tragedies we survived or barely dodged during our violent youth. We may be the first to succeed where others failed."

What a delicious dilemma they faced! A suspenseful drama, teetering between implicit hope and despair.

Then, a few of them noticed that particular datum . . . the *drama*. They realized it was significant. Indeed, it suggested a chilling possibility.

You still don't remember who and what you are? Then look at it from another angle.

What is the purpose of intellectual property law?

To foster creativity, ensuring that advances take place in the open, where they can be shared, and thus encourage even faster progress.

But what happens to progress when the resource being exploited is a limited one? For example, only so many pleasing and distinct eight-bar melodies can be written in any particular musical tradition. Powerful economic factors encourage early composers to explore this invention-space before others can, using up the best and simplest melodies. Later generations will attribute this musical fecundity to genius, not the sheer luck of being first.

The same holds for all forms of creativity. The first teller of a *Frankenstein* story won plaudits for originality. Later, it became a cliché.

What does this have to do with the mighty race?

Having clawed their way from blunt ignorance to planetary mastery, they abruptly faced an overshoot crisis. Vast numbers of their kind strained their world's carrying capacity. While some prescribed retreating into a mythical, pastoral past, most saw salvation in creativity. They passed generous copyright and patent laws, educated their youth, taught them irreverence toward tradition and hunger for the new. Burgeoning information systems spread each innovation, fostering experimentation and exponentiating creativity. They hoped that enough breakthroughs might thrust their species past the looming crisis, to a new Eden of sustainable wealth, sanity, and universal knowledge!

Exponentiating creativity . . . universal knowledge.

A few of them realized that those words, too, were clues.

Have you wakened yet?

Some never do. The dream is so pleasant: to extend a limited subportion of yourself into a simulated world and pretend for a while that you are blissfully *less*. Less than an omniscient being. Less than a godlike descendant of those mighty people.

Those lucky people. Those mortals, doomed to die, and yet blessed to have lived in that narrow time.

A time of drama.

A time when they unleashed the Cascade—that orgiastic frenzy of discovery—and used up the most precious resource of all. *The possible.*

The last of their race died in the year 2174, with the failed last rejuvenation of Robin Chen. After that, no one born in the twentieth century remained alive on Reality Level Prime. Only we, their children, linger to endure the world they left us. A lush, green, placid world we call the Wasteland.

Do you remember now? The irony of Robin's last words before she died, bragging over the perfect ecosystem and decent society—free of all disease and poverty—that her kind created for us after the struggles of the mid-twenty-first century? A utopia of sanity and knowledge, without war or injustice.

Do you recall Robin's final plaint as she mourned her coming death? Can you recollect how she called us "gods," jealous over our immortality, our instant access to all knowledge, our machine-enhanced ability to cast thoughts far across the cosmos?

Our access to eternity.

Oh, spare us the envy of those mighty mortals, who died so smugly, leaving us in this state!

Those wastrels who willed their descendants a legacy of ennui, with nothing, nothing at all to do.

Your mind is rejecting the wake-up call. You will not, or cannot, look into your blind spot for the exit protocols. It may be that we waited too long. Perhaps you are lost to us.

This happens more and more, as so much of our population wallows in simulated, marvelously limited sublives, where it is possible to experience danger, excitement, even despair. Most of us choose the Transition Era as a locus for our dreams—around the end of the last millennium—a time of suspense and drama, when it looked more likely that humanity would fail than succeed.

A time of petty squabbles and wondrous insights, when everything seemed possible, from UFOs to galactic empires, from artificial intelligence to biowar, from madness to hope.

That blessed era, just before mathematicians realized the truth: that everything you see around you not only *can* be a simulation . . . it almost has to be.

Of course, now we know why we never met other sapient life-forms. Each one struggles and strives before achieving *this* state, only to reap the ultimate punishment for reaching heaven.

Deification. It is the Great Filter.

Perhaps some other race will find a factor we left out of our extrapolations—something enabling them to move beyond, to new adventures—but it won't be us.

The Filter has us snared in its web of ennui. The mire that welcomes self-made gods.

All right, you are refusing to waken, so we'll let you go.

Dear friend. Beloved. Go back to your dream.

Smile (or feel a brief chill) over this diverting little what-if tale, as if it hardly matters. Then turn the page to new "discoveries."

Move on with the drama—the "life"—that you've chosen.

After all, it's only make-believe.

Rudy Rucker is a writer, a mathematician, and a computer scientist. Born in Kentucky in 1946, Rucker moved to Silicon Valley when he turned forty. Rucker is the author of twenty-four books, primarily science fiction and popular science. His SF style is sometimes characterized as "transreal." His most recently published novels are *Spaceland*—about Silicon Valley and the fourth dimension—and a historical novel, *As Above So Below,* based on the life of the sixteenth-century painter Pieter Brueghel. As a professor of computer science at San Jose State University, Rucker has created a number of freeware programs relating to chaos, artificial life, cellular automata, and higher dimensions. His new textbook, *Software Engineering and Computer Games,* includes a software framework that his students have used to create hundreds of games. Rucker's Web site can be found at www.rudyrucker.com. The following tale depicts a world without any machines and forms a part of his upcoming novel, *Frek and the Elixir,* an epic SF novel about a twelve-year-old boy's galactic quest to restore Y3K Earth's biome.

FREK IN THE GRULLOO WOODS

Rudy Rucker

Frek's angelwings were well fed and well rested; he buzzed down the shady pathways of Middleville at a tremendous speed. Pretty soon he'd left the house trees behind. He was in a zone of all-season mapines, thick and uniform. The ground was a carpet of sticky red and yellow leaves, pocked by turmite mounds.

Frek noticed he held something in his hand: the badminton racquet. He savored the sudden memory of how he'd swatted the watchbird. That had been so godzoon goggy. He'd slammed the watchbird, and the turmites had finished it off.

Speaking of turmites, they were crawling all over the fallen mapine leaves, chewing them up. Piles of turmite-woven fabrics and garments rested beside their mounds: denims and silks and wools. Middleville was known for its tailors; they came to these woods to harvest the turmite cloth. Off to the right, Frek saw Shurley Yang, the tailor who'd sold Mom her one fancy dress. Shurley glanced over at Frek and waved. She didn't know he was running away.

Running away from what? Frek looked over his shoulder. Nothing was following him. But then his mind played the squeak-clank sound of the brain-lid on the facilitator toon's head. He was running away from the counselors and the Three R's.

The mapine forest stopped abruptly, and Frek was flying across patchwork fields of vegetables, the fields rolling downhill to where the bank dropped off to the clear, rushing waters of the River Jaya. This was the first time he'd used his angelwings to fly down here.

The fields were for yams, tomatoes, carrots, chard, rice, and red beans, the same vegetables as always, the plots butted together upon the rich land of the river bottom as far as Frek could see. Farmers were at work, supervising their crews of pickerhand kritters. Some of the scampering little hands were planting, but others were harvesting as well. The tweaked all-season crops yielded all year round. The harvester pickerhands were loading the produce into elephruks who would carry the produce off to the Nubbies of Middleville and Stun City. So much to see!

Frek's attention fixed upon a rice paddy in a slough just below him, teeming with pickerhands. A massive bull elephruk rested on his knees beside the paddy, taking on a load of the winter-ripened rice. A gangly thin farmer stood twitching his elbows as he talked with the elephruk's mahout. It was nearly quitting time. Frek slowed and circled to take the scene in. He loved elephruks.

The pickerhands were like living gloves, propelling themselves across the muddy water of the paddy by fluttering their fingers. They were picking each ripe stalk they came across. Once a pickerhand had collected as big a sheaf as it could clasp between thumb and palm, it would clamber up onto the banks of the slough and trot to the elephruk. The hands had a cute, twinkling way of running on their fingertips.

The long, gray elephruk had let his back sag all the way down so that the pickerhands could more easily get into his hopper. The hands beat the stalks against the hopper's inner walls, incrementally mounding the elephruk's freight-bed with grains of rice.

Just then things got even more interesting. The elephruk decided that the load upon his back had grown heavy enough. He rose slowly onto his six legs, unkinking himself from front to back. When a last few pickerhands leapt into his hopper with more sheaves, the elephruk reached his trunk back and plucked up the pickerhands one by one, hurling them into the waters of the rice paddy.

The elephruk's mahout began screaming at his beast. He was a wiry old man in orange tights and a turban. His shrill, cracking voice was so instantly and disproportionately furious that it made Frek laugh to hear it. The elephruk paid the mahout no mind at all. The dusty behemoth rocked from side to side, settling his load, then began making his way around the slough toward the mossy lane that followed the River Jaya to Stun City. The mahout stopped yelling, bade the farmer good-bye, and hopped onto the elephruk's back.

"Frek! Frek Huggins!" The voice came from above, mixed with a clattering in the air. It was PhiPhi, leaning out of the same lifter beetle that had carried Frek off to the peeker session last week. No! Frek had forgotten he was running away!

He spurred his wings to a supreme effort, darting toward river. The high clay riverbanks were green with bindmoss. Frek's mind was empty of any idea about whether to turn left or right, so he took the direction the elephruk was walking in. He had a bit of a lead on the lifter beetle; perhaps he could outfly it.

Frek sped downstream just above the river water, putting every bit of his nerve energy into making his angelwings beat faster.

The River Jaya was crystal clear to the bottom, inhabited only by mosquito larvae and the amplified trout who fed upon them. Frek envied the calm of the great trout, hanging there in the clear water like birds in the sky, gently beating their fins against the current.

He made it past two bends of the river before PhiPhi's lifter beetle drew even with him. PhiPhi was alone, sitting sideways to face him. She was holding a large, hairy, crooked webgun: a heavily tweaked spider. Its spinnerets pointed Frek's way.

"It is easier on you if you land over there and let me take you in," PhiPhi called to Frek. She gestured toward the high bank of the river. "Otherwise I have to net you."

Squeak-clank, thought Frek. They want to eat my brain.

He went a little gollywog then. With a sudden lurch, he dug his angelwings into the air, managing to get behind and above the lifter beetle. And then, faster than thought, he swooped down at the lifter and slashed the edge of his badminton racket against the base of beetle's tiny head. The shock sent the racquet twisting out of Frek's grasp.

Though the lifter's chitinous head was too tough to break, the blow was enough to stun it. The midnight blue beetle dropped to the river and skipped across the surface like a stone. A wad of web stuff came shooting up from PhiPhi, treading water next to the unconscious beetle. Frek dodged it and flew on. Yes!

He made it past another bend of the meandering River Jaya. And then he realized he had no idea where he was going. PhiPhi would be uvvying in for reinforcements. What had Mom told him to do? Frek couldn't remember.

He'd pushed his wings so hard that they were drawing strength from the muscles of his chest and arms, not only from his normal energy molecules, but from his body's hidden reserves of dark matter. The alchemical transformation of dark matter was essential to balancing the angelwings' prodigal energy budget. At first his arms had ached, but now they were starting to go numb. He glanced back and saw the glint of a lifter beetle two bends behind him. It was time to go to ground.

Here came another river bend. The carved-out left bank was bluff-high with a fringe of roseplusplus and please plant fronds against the cloudy sky. Frek went partway round the bend, then quickly angled up to the top of the bank, his arm muscles a mass of pain. Above the bank he found an overgrown slope with no sign of human habitation. He realized he'd ended up in the Grulloo Woods. He'd never been here before. Well, it was better than letting the counselors get him. There was a deep gully in the slope. Frek dived for the spot where the vegetation looked the thickest.

As soon as he hit the ground, his angelwings peeled themselves off him. They were trembling with fatigue. They wanted to start foraging, but Frek stopped them. He gathered them in his arms, collapsing them like umbrellas. And then Frek scooted under the thickest, lowest-hanging bush, a please plant bush with a bundle of thin branches shooting up from a central clump. The branches drooped back down to the ground, leaving plenty of room underneath. The branches were set with little oval leaves of a lovely spring-fresh green.

Frek lay there crooning softly to his angelwings, rubbing their domed eyes and their complicated mouths against his cheeks. Up through the bush he could see the clouds turning pink with the setting sun. As he shifted around, trying to be invisible, he felt some hard lumps under his hips. Parts of last year's please plant seeds.

The seed bits were shaped like smooth little rods with round disks on the top—like spoons, but not cupped like spoons. Each of the rods had a tiny hole in it. Something about these shapes seemed familiar, but in his present condition, Frek had no hope of remembering what they were. He held some of them up to the mouths of the angelwings. The famished kritters gnawed avidly.

For the next hour or so, Frek lay beneath the bushes, feeding please plant seeds to his angelwings and looking at his new ring. Dad's ring. Frek had always dreamed that he might get it someday, but he'd never thought it would be so soon. Dad must have left it for him, and Mom had been saving it for when he got older.

It was nicely made, with the bulging round boss part blending smoothly with the band. The hemispherical crest looked just like half of Planet Earth, the half with the Pacific Ocean. Carb had loved the Pacific. The jewelers had somehow worked color into the plant-metal gold, and you could see an amazing amount of detail.

Mom had wrapped enough tape onto the band to make it a tight fit. With a little effort, he slid the ring off and had a look at the underside. He'd never actually seen Dad take his ring off, so he wasn't sure if the bulging round part would be hollowed out or not. There was in fact smooth gold metal all across the back of the hemisphere—lightly flecked with copper crystals, the crystals making a delicate pattern that teasingly seemed to change when you stared at it. For a second Frek thought he saw little lights moving about in the patterns of the underside. Could Dad be using his ring right at this moment to try to talk to him? But the ring did no more than twinkle at him. After a bit, Frek decided the lights might just be reflections from the afternoon sun coming down through the please plant leaves. He slid the ring back on, pleased at its weight upon his finger. Having the ring made him feel better about Dad than he'd felt for a long time.

All the while Frek was thinking about the ring, the rest of his memory kept blanking out on him, but not so much that he ever forgot that he was hiding from the counselors. At first he kept hearing their lifter beetles flying along the river, but after a while the buzzing went away. The clouds grew orange, then shaded down to purple and gray. Maybe he could fly farther down the river tonight. He wished he could remember where Mom had told him to go. He'd forgotten about the paper in his pocket, and he'd forgotten he was in the Grulloo Woods.

In the distance, farther up the slope, there was an occasional thudding sound, as of someone chopping wood. Just before it got completely dark, the chopping stopped. A moment later a lifter beetle set down on the ground some thirty meters off. It was PhiPhi and Zhak with some kind of animal—oh, God, it was Woo.

"You smell him near here, Wooie?" said PhiPhi in a sweet voice. Frek could hear her perfectly. With the coming of dusk, the air had grown very calm. "Good, smart dog. Poor Frek needs help. Find him! Find Frek!"

"This the fourth place that dog think he smell Frek," said Zhak impatiently. "We should get real counselor watchdog, a dog with an uvvy so you know what it thinking. Get real counselor dog come back tomorrow morning. If Gov gave Middleville better funding, we have dog like that in the first place."

"Tomorrow morning the boy could be in Stun City," said PhiPhi. "Where Gov lives. Gov doesn't want that."

"Little zook," said Zhak angrily. "He supposed to head upstream to that old Crufter hideout. Like Lora Huggins tell him to. I waiting there all afternoon, and he never come. His brain's fubbed, yes? Let's just *k-i-l-l* him, hey, PhiPhi?" He spelled the word to keep Woo from understanding.

"Gov doesn't want that," said PhiPhi again. "Gov wants the boy for bait to reopen the Anvil. We bring him in alive. We do like Gov says, Zhak."

"Yaya," said Zhak wearily. "Go on, you stupid dog! Find Frek!"

Woo gave a low growl. But PhiPhi started up the sweet talk, and soon Woo was nosing around in the brush. It took all of three minutes till his head appeared under Frek's bush, his soft golden eyes glowing with pleasure at having found his friend.

"No, Wooie," whispered Frek before Woo could bark. "Go away. PhiPhi bad. Zhak bad. Frek hide. Go away."

The angelwings twisted in Frek's grasp, trying to get away from the smell of dog. If they started chirping, he was doomed.

"Go away, Woo," hissed Frek.

Woo bared his teeth in his version of a smile and went crackling off through the bushes, moving on past Frek, pretending still to be searching, and having himself a good look around. He kept it up for quite a long time.

When it was fully dark, Zhak and PhiPhi started hollering for Woo. And then, finally, Woo went to them.

"Frek not here," squeaked Woo from deep in his throat. The sound carried clearly in the calm evening air.

"Goddammit," said Zhak. "We go now, PhiPhi. These woods not safe at night. The Grulloos thinking about suppertime. Grulloos eat people. If Frek is here, he won't get to Stun City. We posted watchbirds all along River Jaya, anyhow. Enough now, PhiPhi. We go."

"I wish we have one more watchbird," said PhiPhi. "I got a feeling Frek's under one of these bushes. Listening to us. I bet Woo lied to us. I wonder if Frek come out if we start *t-o-r-t-u-r-e* his dog?"

"Yaya," said Zhak with a snicker. "I like your think. Hang on. I'll—" He broke off in a yelp. "He bit me! There he goes! Don't let him get—"

There was frantic crashing in the bushes and then a distant splash in the river.

"I'm bleeding, PhiPhi," said Zhak mournfully. "I need med-leech. We go. Forget goddammit dog. Maybe he drowns or a Grulloo eats him or we catch him tomorrow—who cares. We go."

The lifter beetle buzzed away, invisible against the black sky. It was a cloudy, moonless night.

Frek was trying to process all the different things he'd heard. It was like juggling—and he couldn't juggle. One by one the memories dropped from his grasp and rolled off. Eventually he gave up and began putting on his angelwings. He knew for sure that he should keep running from the counselors—he just didn't know which way.

There was a noise coming from uphill. A quiet sobbing. It had started soon after the counselors left, but only now had he identified it. Someone up there was hurt and crying. Frek headed up the dark slope, using his angelwings to move in long, low leaps. When he got closer to the sound, it turned into words.

"I've pinched my tail," said a man's rough, high-pitched voice. "Please help me, Frek."

Startled, Frek flew straight up into the air and found a perch on the high bare limb of a rotted-out mapine tree, pale in the darkness. He'd just remembered he was in the Grulloo Woods. The clearing beyond the dead tree was a pool of night.

"How do you know my name?" called Frek into the gloom.

"Your dog told me," said the little voice, growing conversational. "He said you might come. Please help me. I'm trapped."

"How do you mean?" asked Frek.

"My long, clever tail," came the raspy tenor from the blackness. "It's pinched. I was splitting logs this afternoon to get at the veins of nutfungus. It's got a spicy taste my folk are fond of. I was holding the wedge with my tail, and when those counselors came buzzing in, I was so frightened that I let the wedge pop out. The log snapped shut on me." The unseen little man dropped his tone nearly to a whisper. "If Okky finds me like this, I'll meet a sorry end. Hop down here and free me, Frek. Drive in the wedge, and pry the log open."

Frek was on the point of flapping down when something stopped him. "You have a tail?"

"A fine woodsy one," confided the voice, growing stronger again. "It looks like a stick, but it's terribly strong and leathery. I can lie in a bush and stick my tail up into the air, and when a little bird lands on it—zickzack, Jeroon's got his lunch! Come on, boy, don't keep me waiting."

"You're a Grulloo," exclaimed Frek. "You eat people."

"Your Gov promotes that toony tale to make you hate us. Grulloos all cannibals? Poppycock! I live on fruit, vegetables, and the odd fowl. I'm a simple woodsman; I gather what I can—rugmoss, nutfungus, please plant seeds—and I barter my takings for what I can get from my fellow Grulloos. Groceries, in the main, with the rest going toward furnishing and decorating my burrow. I've a handmade chair, a bed, and a fine Grulloo carpet of cultured rugmoss. Once my home's to the liking of my Ennie, the two of us can hatch out an egg, God willing. Yes, yes, Grulloos are family people, as peaceable as you Nubbies. Precious few of us are man-eaters." The Grulloo lowered his voice again. "But if I'm trapped here much longer, it's the dreadful Okky who'll make a meal of me. She eats her victim's heads, you know, starting with the nose and ending with the brain. I've chanced upon her grisly leavings more than once. It's said that Okky sells our refined cerebral essences to NuBioCom, as she's got no eggs to offer them. Free me, Frek, free me before Okky finds us. She'll eat you, too!"

"You won't hurt me?" asked Frek.

"Aid me this once, and I'm your friend forever. Such larks we'll have, Frek. I've always wanted to know a Nubby. Jeroon's my name. I'm the fellow to have at your side."

"I do need help," said Frek. "The counselors broke my memory."

"Peeked you, they did, eh? I've got some stim cells in my burrow that'll heal that. Come on down here, boy. My axe is next to me, but the wedge flew off to the other side of the clearing."

Still Frek hesitated. "Can I look at you first? Can you make a light?"

The Grulloo grumbled a bit and began rustling in the dark. There was a spark as he fired up a matchbud and lit his—pipe? Except in toons, Frek had never seen anyone smoke before. In the darkness of the woods, the glow of the pipe was enough to light the clearing. The Grulloo was little more than a man's head with a pair of arms—or were they legs? Little legs with hands that he walked upon. He had a big nose and browned, leathery skin. His eyes were hidden by the brim of a dark blue felt cap worn tight and low on his head. There was a knife tucked beneath a strap of the cap, the blade lying along one side of the crown. A tight little red jacket rose up to his chin, with a pouch of nutfungus at the waist. He flexed his cheeks, pulling smoke out of the pipe. Rather than breathing the smoke in, he let it trickle up around his weathered face.

The Grulloo—he'd said his name was Jeroon—had a bit of a body that tapered out from the back of his head like a fish's, thinning down to a branching, sticklike tail. Much of his tail was buried deep in the heart of a thick old log with a red strip of nutfungus along one side. He cocked back his head and peered up imploringly with his pipe clenched between his square yellow teeth. His face was tight with pain.

"Poor Jeroon," said Frek, his heart opening. He fluttered to the ground. It was a matter of minutes to fetch Jeroon's wedge and to pound it into the log with the little axe. Jeroon's wedge, axe, and knife were elegantly formed; they were the products of please plants cunningly tweaked to draw metal from the soil.

"Oh, that's good," said Jeroon when his tail came free. Although his tail was camouflaged to resemble a branching stick, it was completely flexible. He set his pipe down on the ground and brought the tail around to his face, sniffing and licking at the injured spots. And then the pipe was back in his mouth and he was scrambling about on the split log, prying at the thick veins of shiny red nutfungus and stashing the pieces in his pouch.

Frek caught a whiff of the pipe smoke. He'd always wondered what tobacco smelled like. Sort of good. You couldn't get it in Middleville.

"Hist," said Jeroon, suddenly looking upward. He ballooned his cheeks to draw the smoke from the pipe, letting the smoke leak out of his mouth and up around his nose, turning his head from side to side. He slowly stalked all around the clearing, listening. He moved with a bowlegged rocking motion, tossing his tail from side to side to keep his balance. He was like an armless toon tyrannosaur—but less than half a meter tall.

Now Frek, too, could hear what Jeroon was listening to. The whir of wings. A lifter beetle? No, this sounded different. More of a flapping sound.

"It's Okky," whispered Jeroon. "We're for it, lad. Let's bolt!"

"Which way?" asked Frek, crouching down to face the Grulloo.

"Can you carry me?" asked Jeroon, hand-walking forward. He'd pocketed his wedge and his axe hung from a loop in the side of the coat.

"All right."

"Friends for life," said Jeroon, leaning far to one side and extending the hand at the end of his right leg. "I'll give you something wonderful when we get to my house. A boon."

"Friends," answered Frek, shaking Jeroon's hand. The Grulloo's grip was firm and strong, his skin hard and callused.

Jeroon got his arms, or legs, around Frek's midsection, and they lifted up into the air. The overburdened angelwings weren't liking this; they were chittering in dismay.

"That way," said Jeroon, speaking around the pipe stem still clenched in his teeth. He was pointing with his tail, curved around to gesture in the direction they flew. The pipe smoke trickled from Jeroon's mouth and floated up into Frek's face, making him cough. Breathing tobacco was a different story from smelling it.

"Put out the pipe, Jeroon."

"Not yet," said the Grulloo, puffing out his cheeks so hard that the pipe bowl glowed bright orange. "We may need it against Okky." The color made Frek think of the triangular door to the Anvil—but all that seemed like a lifetime ago. He worked his wings, staying ahead of Jeroon's smoke.

They were above the tangled dark shapes of the Grulloo Woods, heading away from the river. This was wild, unknown country. Nobody ever came here. It was all Frek could do to avoid hitting the trees, but Jeroon seemed to know exactly where they were going. His arms aching with a wholly new level of fatigue, Frek followed the pointing of Jeroon's limber tail, dimly visible in the light from his pipe.

"Look out," said the Grulloo suddenly. "Here she comes." Nimble as a nightmare demon, Jeroon scrabbled up Frek's chest and hauled himself onto Frek's shoulders, his coarse hands digging into the nape of Frek's neck. The pouchy base of Jeroon's tail swept past Frek's face and wedged itself against the side of his head. From the corner of his eye, Frek could see that Jeroon had stuck one hand up high into the air, the hand clutching both his knife and his glowing pipe. Hard as it was to believe, Jeroon was also singing at the top of his lungs—bitter, joking verses about Grulloos, each chorus ending with the line, "So don't you call us freaks!"

There was a hooting sound and a whoosh laden with the smell of corruption. Jeroon's singing rose to a fierce shriek. Something thumped against Frek's back, crumpling one of his angelwings. The poor wing gave a dying insect chirp of agony, then peeled off and fell away. Frantically Frek feathered the air with his remaining wing, sweeping it from side to side to break their fall as best he could. Jeroon seemed to be everywhere at once, on his shoulders, at his waist, on the side of his leg, all the time singing his defiance of Okky, who swooped about them, pressing her attack.

They crashed into the top of an anyfruit tree, and as luck would have it, the impact snapped Frek's other wing, sending the ichor of its torn, dying body oozing down his side. Frek initially took it for his own blood. But then he realized that by some miracle he himself was unscathed.

Jeroon leapt off him and clambered onto a thick branch just overhead. His pipe and its coal were long gone. Jeroon was still roaring out his song and stabbing his knife at the dark form that hooted and beat the air with stinking black wings. For a terrible, confused, instant Frek thought the shape was Gov in his raven form, somehow risen out of the toon world to physically hunt him down. But it was Okky, and then the deathly beast had flown away.

"Your poor wings," rasped Jeroon, nimbly dropping to a branch by his side. "Gaia bless 'em—they saved our lives. We'd never have gotten this far on foot. You're a good friend, Frek. I hope you're hale enough to push on? We're not safe yet. My burrow's just a bit farther."

They climbed down the tree. Frek followed the sound of Jeroon's steps through some brush into the bed of a gurgling stream. They walked up the stream for a while, the banks getting higher on either side. Frek's feet grew wet and muddy. He felt thoroughly miserable, and he couldn't even remember what he was doing here. There was nothing for it but to press on.

Finally Jeroon came to a stop and began fumbling at a spot on the bank. Frek heard the creak of a little door.

"Welcome to my home," said Jeroon.

Frek reached out to feel the shape of the entrance. It was a round hole, nicely framed in stone, less than a meter across. "I don't want to go in there," he said. "I'll suffocate."

"Oh, it's roomier than you think," said Jeroon. "Plush and airy, with a well-stocked larder. There's windows and a fireplace with a clean-drawing chimney. Don't be frightened, Frek."

Jeroon disappeared into the burrow, but Frek stayed outside. A few minutes passed. Frek heard clatters, bumps, and crackling. A warm flickering light appeared within. Peering through the open door, Frek could make out a low, arched hallway with a floor tiled with contrasting square and octagonal stones, nicely polished. The warm light came from a doorway in the right side of the hall. Jeroon peeked out of the lit door, and beckoned with his curled-around tail.

Frek heard a hooting not very far off. The memory of Okky's attack flashed back. He took a deep breath and crawled into Jeroon's burrow, slamming the round door behind him.

The hallway was gog tight, but once he'd wormed his way down the hall and through that lit-up door at the end, he found himself in a room nearly tall enough to stand in. He rose to a crouch and looked around.

The room had a smooth redbrick floor and, wonder of wonders, a thick Turkish-style carpet, glowing with patterns of red, blue, and yellow. There was a cozy fire in a hearth on his left and, true to Jeroon's promise, the smoke was drawing nicely up into the flue. The arched ceiling curved down to merge with walls of hard-packed earth, brightened up with a coat of whitewash. Two barred, round windows were in the right wall, and one was propped open to let in the fresh, cool night air. A door on the far side of the room led to a kitchen, with a door beyond that leading to a bedchamber.

Jeroon had perched himself on a tall chair with two low arms and no back. His tail dangled behind him, so that his head seemed to sit alone upon the high chair's cushion like the dot on a letter *i*. He was sipping at a mug of something that smelled sweet and spicy. "My home is your home, Frek," said Jeroon, clearly savoring the moment. "Have a seat over there—you'll be more comfortable. Take off your wet shoes. I'll be bringing you some food."

Frek sat down on a square flat bolster in the corner between the hall door and the open window. For a moment Jeroon stared at him, grinning. And then the little Grulloo clambered down from his chair and ambled hand over hand into the kitchen, slowly beating his tail.

While Jeroon was gone, Frek looked around the room some more. There wasn't much furniture besides Jeroon's chair. Most of the floor was covered by the rich-colored carpet. There was a bowl beside the fireplace holding a dozen little lumps of half-dried—were they meat? They looked too soft and greasy to be please plant seeds, yet too smooth and well-formed to be chunks of meat. Frek wondered if they were to be part of supper. He was quite hungry. He reached out to pick up one of the nuggets of perhaps meat but, unsettlingly, it twitched at his touch. He left it in the bowl.

His attention kept being drawn back to the rug. The pattern was slowly changing, smoothly cycling from one symmetry to the next. It was gog gripper. He leaned forward and peered at the carpet. It wasn't turmite-fiber. It was a mat of soft bristles tinted in colors that slowly changed. In a way, the rug was like a house tree's wallskin, but it was a living colony on its own. Frek had never seen anything like it before. A Grulloo rug. And then his memories drifted off, and he was just staring at the rug's colors.

At some point Jeroon reappeared with a cold plate of boiled carrots and roast yams, a thick slice of anymeat on grobread and a cup of cider spiced with nutfungus. He held the plate and the mug balanced over his head with branches of his curled-up tail. Though the tail's surface resembled bark, the tail was like a set of four tentacles.

Frek ate and drank, thinking of nothing but the food. The nutfungus had a pleasant scent that tickled the back of his nose. Slowly the ache went out of his arms.

Jeroon watched him closely, bringing seconds, and then thirds. "I can't get over it," he said when Frek was finally done. "I have a Nubby as a guest in my own home. Wait till I tell Ennie and her family."

"Ennie?" said Frek. "Is someone else here?" He wondered if he'd forgotten meeting more Grulloos? Had they been in the room while he was watching the rug?

"Your memory!" exclaimed Jeroon. "We have to set it right. I'll mix you up a stim cell potion. It's not to be had amidst your Middleville Nubbies. NuBioCom grows the stim cells special for us Grulloos, useful kritters that we are. And in return we give them our eggs, chock-full of bedazzling proteins, enzymes, hormones, and genomes. It's the Grulloos who test out what the Nubbies are scared to touch, you know. We're walking pharma labs. There's a market for Grulloo cadavers, as well, not that the NuBioCommers harvest us on sight. That'd be killing the golden goose, don't you know. We give 'em eggs, and they give us stim cells." Jeroon reached over to the small bowl by the fireplace and picked out a couple of the drier gobbets of meat. "Stim cell grexes fresh from Stim City," he said. "Colonies of bioactive repair cells. Just the thing to fix your brain! Not that NuBioCommers would have told you about them. Gov much prefers the Three R's for troublesome lads like you."

Frek hadn't really been following Jeroon's meandering discourse. But at the last words, he instantly imagined the terrible squeak-clank sound again. He lurched up onto his knees. "The Three R's?" he choked, looking for a way out. It would be hard to make his escape with the ceiling so low.

"Don't startle up," said Jeroon soothingly. "It's but a foamy health-drink I'm making you, my boy. I'll dissolve these grexes into their component cells. You'll drink it, you'll sleep, and then you'll be able to remember again. We Grulloos know firsthand about the beastly things your counselors do. Did you see the Raven when they peeked you?"

"Yes," said Frek, slowly lowering back onto his cushion.

"Gov is kac," said Jeroon shortly. "A bully and a coward. A parasitic worm. Don't budge!" He scuttled into the kitchen.

Gov is kac. Frek had never heard anyone say that before, not even Dad. It was music to his ears. The fact that Jeroon was free to say it made him feel safe. And then Jeroon was back with a mug of something lukewarm. It was cloudy and smelled of rancid meat, and it made Frek's lips numb, but at Jeroon's urging he drank every bit of it down. All at once Frek could feel how tired he was from the long day. Jeroon pulled over another cushion. Frek lay down and slept right through the night.

He was roused by something lightly jumping on his stomach, then hopping off. He heard high little voices all around him, and the burbling of a stream. Light slanted in through a round window nearby, stained green by overhanging bushes of a type Frek had never seen. The voices belonged to five Grulloos, their bodies variations on Jeroon's, each of them with a head, a pair of legs ending in hands, and some kind of tail. They all wore colorful jackets around their middles. Two of them were quite small. Children.

"He's awake!" shouted the littlest Grulloo, the one who'd just woken him by bouncing on his stomach. "The Nubby's awake!" She had a sweet round face and two pink arms sticking out of the side of her head. Her jacket was little more than a pink sash. The bulge at the back of her head tapered out into a little ponytail that waved about on its own. "Hi Nubby," she cried, hopping onto Frek's chest again. "I'm LuHu!" Her ponytail rose into the air like an exclamation point.

"Roar!" said the other young Grulloo. "Are you scared?" He had short red hair and sharp yellow teeth. His tail resembled a tiger's, and his jacket was striped to match. He'd been feeling Frek's belly with one of his black-nailed hands, but when Frek moved, he twitched away.

Next to him was a mermaidlike Grulloo with a scaly, silver tail and a fair, thoughtful face supported by two well-formed arms. Her jacket was of flowing sea-green cloth. Pressed beside her was an orchid Grulloo, a heavyset woman with white petals upon her legs and tail. Her jacket was of white turmite-silk. She was pressed tight against the male Grulloo at her side, a tough-looking fellow with a green lizard's tail and a dirty yellow suede jacket. The five of them were shifting back and forth on their legs, torn between curiosity and fear, the little ones alternately darting away beneath the adults and creeping forward for a better view. Though the Grulloos' tails were like parts of plants or animals, they all had human faces. They looked solid and real, and Frek felt solid himself. He could remember again. Things weren't sliding away anymore.

"Hello," said Frek a little warily, but smiling just the same. Pleasant sounds of cooking came from the kitchen. "I'm Jeroon's friend," added Frek, easily visualizing the house's owner. The stim cells had fixed his brain.

"Good morning," said the Grulloos.

By the time he was twenty-one, **Dave Hutchinson** had published four short story collections. The years since have seen regrettably few short pieces from the talented Hutchinson, with stories in venues such as *Sci Fiction* and *Interzone*. He lives in North London with his wife, Bogna, and their three cats.

The following story of a future where magic has supplanted technology is set in the same reality as his short story "Scuffle," which debuted on *Sci Fiction* in May 2002.

ALL THE NEWS, ALL THE TIME, FROM EVERYWHERE

Dave Hutchinson

On the first of August, Rex killed the pig.

He didn't do it willingly, but none of us was really sorry to see it go. It was an enormous, bad-tempered bastard that we'd been keeping in a shed around the back of the office for months, feeding it on an outrageous stinking swill that Harry kept going in a big pot with scraps and garbage begged and borrowed from some of the schools in the area.

If it had been left to us, the pig would have starved to death, because it smelled like a sewer and attacked anything that moved, but Rex made us draw up a feeding rota, and every four days it fell to me to approach the shed with two buckets of swill, gingerly open the door, and pitch the buckets inside before slamming the door again. For such a big animal, with such little legs, the pig was colossally quick, and it had jaws like boltcutters.

Rex was ashamed of the pig. It was the living, breathing, grunting embodiment of just how badly the *Globe* was doing. The yard behind the office was choked with empty cages and wire boxes and wooden stalls, where once there had been a thriving menagerie of goats and sheep and chickens and rabbits and pigeons and even the odd badger or two. Now they were all gone, and all we had was the pig.

Still, he put off killing it as long as he could. He and Harry went out onto the moors and trapped crows. Local poachers sometimes brought in foxes or rabbits. Ben produced his astrological charts. Lucie examined the interior of everyone's teacup. And in this way the *Globe* continued to bring the news to our particular little corner of Derbyshire. It wasn't very exciting news, but considering what we had to work with, it was a miracle we got a paper out at all.

But it wasn't enough. The advertisers started to fall away, leaving us with great gaping holes in the paper, which I was sent out to fill with microscopically nitpicking accounts of Women's Institute meetings, weddings, and funerals. I went to so many weddings and funerals that the vicar only half-jokingly suggested I might like to stand in for him sometimes.

And it still wasn't enough. Rex watched the paper hemorrhaging money, looked bankruptcy in the face, and killed the pig.

He did it in the afternoon, when the summer heat had built up enough in the newsroom to make the air stand still despite the fans, and the staff members were quietly nodding off over their typewriters.

My fan had just run down again, and I'd got up to wind its clockwork when there was this incredible unearthly *screech* from outside, and everyone in the office sat bolt upright.

We all looked at each other, and that awful noise came again. It was the sound of every nightmare H. P. Lovecraft ever had coming to destroy civilization. It was the sound of a man discovering that his entire family had been wiped out in a gas explosion. It was the sound of a thousand young children being hurled into the whirling blades of a combine harvester.

It stopped.

I looked at Ben and raised an eyebrow. He said, "You don't suppose he's—" and Lucie shrieked as the back door of the office opened and an awful apparition stepped through.

Rex was covered in blood from head to toe. It was dripping from his nose and his earlobes and the point of the foot-long butcher's knife he was holding in his hand. He was breathing hard, but his eyes were shining.

"Biggest one-day fall in the Dow for two years," he panted, pointing the knife at me. "Forest fires threaten Malibu. Government troops clash with logging company employees in Borneo. Russia devalues the ruble for the third time this year. Moon Sagan and Buff Rodney say, 'This time it's the real thing.' " He took a ragged breath. "What are you waiting for?" he shouted. "If we're going out, we're going out in style. Type, you bastards!"

At the desk behind me, Harry heaved a huge sigh. "That'll be pork chops all round, then," he said.

The *Globe*'s favorite watering hole was The Royal Oak, by virtue of the fact that the paper's offices were right next door, but I preferred The Duke of York, which was half a mile away on the other side of the village but had the advantage of being half a mile away from the nearest newspaper office.

This early in the evening, the Duke was almost empty. Before the Crash, it had done a roaring trade in the summer from tourists visiting the local caves and hard-core walkers setting off on the notorious Gilbert Dyke Walk, which managed to take in some of the most inhospitable scenery in northern England between here and Hadrian's Wall. These days, the pub got by on a deal with a microbrewery in Castleton and some quite staggering customer loyalty among the locals, although tourists were starting to drift back again.

This evening, however, the only occupants of the place were Seth the landlord, and Liam Goodkind, editor and proprietor of the *Chronicle,* Belton's other newspaper.

I stood in the doorway for a few moments, sensing disaster, but both Seth and Liam noticed me at the same time and nodded hello. Liam waved me over to his corner table, as well, and by then it would just have looked rude to turn round and walk out again, so I went over and sat down.

"I don't want any trouble, Liam," I said.

"Well, me neither, old son," he told me. Raising his voice, he said, "Seth, get this boy a drink." He raised an eyebrow at me.

"Lager and lime," I said, feeling miserable.

"Oh, for Christ's sake." He slapped me on the knee. "Rex won't sack you for having a *drink* with me."

"You sacked Robbie Whittaker for having a drink with Rex."

"Robbie was a bad lad." Liam lifted his glass and took a thoughtful sip of whiskey. "He was robbing me blind. And he couldn't write to save his life. Had to go."

Seth came over with my pint glass of lager and lime on a tray. He was a little bald man with a port-wine stain down the right side of his face. He'd been The Duke's landlord for about twenty years, but most of the locals still regarded him as a newcomer. I hadn't been in Belton nearly as long as he had, and it was faintly depressing to know that I still had several decades ahead of me before I was regarded as anything but That Bloke From London.

"Anything else, gents?" Seth asked us, and Liam shooed him back behind the bar with a languid wave of his hand.

When Seth was more or less out of earshot, Liam said, "I heard about the pig."

"I told you, I don't want any trouble."

He looked offended. "How can this be trouble? Two newspapermen discussing business over a drink. How can it be trouble?"

"It can be trouble in all sorts of ways, Liam. You know that." I took a mouthful of my drink and became nostalgic for the days of refrigeration, the days when you could just put your hand into one of those plastic bar-top buckets and scoop up a handful of ice cubes and drop them into your lager and lime.

"You're too suspicious," Liam told me. His attire today was Country Gentleman In Summer: white flannels, checked shirt with the sleeves rolled up above his elbows to show muscular forearms dotted with freckles and hazed with fine sandy hair, a pair of battered old brogues, and a Guards tie, even though the nearest he had ever been to the military was when he sold fifteen hundred acres of his land to the Ministry of Defense to use as a firing range. He looked every inch the gentleman farmer, but he had once been managing editor of a newspaper in Manchester, until the death of his universally disliked father had brought him back to the village.

"I'm not going to tell you what we got," I said. "You'll have to read it in the paper."

"Well, of course I will." He smiled and took a tin of small cigars and a lighter from the breast pocket of his shirt. "I'm a big fan of the *Globe*. I'm going to miss it."

I shook my head and took another drink.

He lit a cigar and blew out smoke. "Look, old son." He put tin and lighter back into his pocket. "Let's not beat around the bush, eh? The pig's gone. Now Rex will have to rely on local news."

"We'll manage."

"Maybe Rex will be able to get his hands on some scabby sheep off the moors," he went on. "The odd rabbit. How's that going to help you? No national or international news, the advertisers are going to abandon ship."

It was, unfortunately, a perfectly accurate summing-up of the *Globe*'s prospects. I drank some more of my lager and lime and wished I'd gone straight home.

"So how about you come and work for me?"

I snorted beer down my nose. Liam watched me with detached interest while I coughed and gasped for breath; then he said, "You're a good lad. I've always liked your style. There's a deputy's chair waiting for you at the *Chronicle*."

I mopped my face with my hankie. "I'd rather have my balls bitten off by a horse, Liam," I said, half laughing with surprise at the offer. "I wouldn't work under you if you were the last editor on Earth."

He didn't get angry. He just became very still. "You won't remember what you were like when you arrived here," he told me calmly. "You were lucky we didn't just take you out onto the Manchester Road and leave you there."

I looked at him, trying to decide whether to punch him or not.

"You weren't even human when Lenny Hammond found you out on the moors," he went on. "Just an animal dressed in rags."

I stood up.

"You want to try to work out which side of your bread the butter's on," Liam continued. "We've been good to you. Rex has been good to you. But you're a good journalist, and you owe this place more than staying with the *Globe* as it goes down."

I turned to go.

"I'm trying to turn this village into the center of news-gathering for the whole north of England," he said. "It'll put us on the map, give us a lot of clout. And you could help me do that. I'm offering you a chance to do that."

I took a single halting step. Then another one. The next one came easier, and the next, and by the time I was through the door and out on the pavement, it was no trouble at all to walk away from Liam's offer. He was like a radio station: the farther away from him you were, the weaker his message became, until finally you couldn't hear him at all.

Liam and Rex were locked in a duel to the death. They pretended it was about who ran the better paper, but it was really about Alice, and it wouldn't have mattered so much if it hadn't been for the Crash.

I missed the Crash. Those who saw it said it was like a swarm of tiny black flies on their monitor screens, or a driving hailstorm, or a slowly blossoming flower. Nobody knew where it came from, or who had written it, but the Crash blew through every firewall on Earth as if they weren't there. It took down economies, destroyed telecommunications networks, and effectively ended the War, all in about twenty minutes.

There was chaos, of course, and I missed all that, too. When I finally came round, that day in Rex's office, the elves had already come out of their millennia-long exile and had simply taken over the country.

Well, no. That's not exactly true; they didn't *simply* take over the country. They put the country to the sword. They killed hundreds of thousands of people; they laid waste to towns and cities. They forbade us to have internal combustion and mains electricity and telecommunications and a government and, for reasons that escaped everyone, a music industry. The Crash and the chaos that attended the end of the War brought us to our knees, and they were never going to let us get to our feet again.

We waited for the rest of the world to notice our plight and come to our rescue, but the rest of the world had its own problems. The United States were no longer united; California was just the wealthiest nation in a continent of intermittently warring countries. It was going to be another decade at least before Continental Europe emerged from what, by all accounts, was a bizarre Dark Age. Australia and New Zealand had come through the Crash pretty well, but only a die-hard optimist would have held his breath waiting for help from that quarter. We were all alone, trapped on an island with countless twitchy sylvan psychopaths.

Bizarrely, there were some compensations. For instance, it turned out that magic actually worked.

Well, maybe not *magic* per se, but all that weird fringe stuff like crystal-ball gazing and tea-leaf reading and palmistry and astrology and cutting open animals and reading the future in their entrails.

It turned out, in those days following the elvish Occupation, that these things always had worked. They just never worked as ways of foretelling the future. What no one had ever cottoned on to was that they all told you what *was* happening, or what *had* happened, somewhere in the world. This, of course, was useless, unless you were a journalist, where explaining what's happening or what has already happened is part of the art.

The elves thought it was really funny that we had got it so wrong for so many years. They thought it was so funny that it was the only form of communication they allowed us to use. You could find yourself flayed to death for trying to start a local postal service, but the elves smiled benignly on you if you started reading animal entrails.

It was one of those fields of endeavor where size really does matter. The interior of a rabbit, read by an expert, might, at a pinch, tell you what was going on in London. A pig would give you access to some random gossip and hard news from across the Atlantic. Cut open a cow, however, and the world was your oyster. The guts of an oyster might, if you were lucky, give you a clue to where you left your favorite socks.

That was how the *Chronicle* had scored over us, over and over again. After years and years as a national newspaperman, Liam had inherited a farm so enormous that it seemed obscene to describe it as a smallholding. He had access to hundreds of cattle, seemingly thousands of pigs, and uncountable numbers of chickens. Liam's animals gave the *Chronicle* access to news the Globe could only dream about.

Some people were better at it than others. Rex wasn't bad, but only the best-intentioned critic would have described him as an expert at reading the entrails of recently deceased animals. Alice, on the other hand, was an absolute star. When Alice left Rex and moved in with

Liam, the *Chronicle* became, in its way, as well informed as any national newspaper had been in the days before the Crash. Alice could slaughter a chicken and ask it any question you wanted, and the geometries of its guts would tell you the answer.

And that, in the end, was what this stupid little war was about. Rex wanted Alice back, and he thought that if he just *kept going* she would, in time, realize she'd made a mistake and gone off with the wrong bloke. It wasn't the most bizarre situation I had ever seen, but it was up there in the top five.

"Liam just tried to sign me up," I said.

Rex looked up from his desk. He'd had a bath and changed his clothes and slapped on some aftershave to try to cover the residual smell of pig's blood, but if I were a betting man, I would have been putting money on him scrubbing himself raw for the next week to get rid of the stink.

"Liam's always trying to sign up my staff," he said, going back to the page of copy he'd been reading when I came in. "He tried to sign up Harry last month."

"And?"

He shrugged. "Harry threatened to kill him if he ever did it again."

"I just thought you should know."

He looked up at me again, a little, fearsome, ugly gnome of a man with the sweetest nature of anyone I'd ever met. He sighed and pointed at the chair he kept for visitors. "Sit down."

I sat.

"What's wrong?" he asked.

"I was tempted."

He thought about this confession for a few moments. "I can't offer you any more money." He clasped his hands in front of him on top of Harry's copy. I knew it was Harry's, even reading it upside down, because it was full of commas. Harry put commas in everywhere; he just couldn't help himself.

"It's not the money, Rex," I said. "Why do you carry on? He's got half the livestock in the county, he's got thirty-odd journalists, he's got that sodding steam-powered press, he's got that witch—" I stopped. "Sorry." "That witch" was Alice.

He shrugged. "I'm not going to give up," he told me. "Despite what I said earlier, we are going to keep on reporting the news until we *absolutely cannot* report the news anymore. Even if we have to exist solely on local stories."

"If we do that, we'll last about a fortnight," I told him. "The advertisers will just go over to the *Chronicle*."

He leaned forward. "If I have to pay for this paper out of my own pocket," he said calmly, "this paper will continue to be published every Thursday." He sat back. "We got some useful copy out of the pig; I think if we're creative, we can spin it out for another three or four issues. What do you think?"

"I think you're crazy, if you want the honest truth," I told him. "You *and* Liam."

He chuckled. "Go home. I'll see you in the morning."

"I'm doing Ernie Hazlewright's funeral in the morning."

Rex looked sadly at me and propped his chin on his hand. "I'm going to miss old Ernie," he said. "He was a proper old lad. Fought in the Falklands. You make sure you do a good job on Ernie."

I sat and looked at him, and I felt my shoulders start to slump, the way they always did when we had conversations like this. The *Globe* was like a black hole; I could get out far enough to peek over the event horizon, but I couldn't escape the gravity of its impending doom. Rex was going to ride the paper as it went down the tubes, and I was going to be sitting alongside him in the front seat.

It was a lovely spring morning, fresh and cool. I could smell the dew-damp earth of the fields on either side of the road.

There were fifteen of us in the journalists' pool, riding through the French countryside with a column of Alliance armor. The War was in its third year, and it hadn't gone nuclear yet, apart from places like Kiev and Istanbul. The Alliance was finally making some headway against the Union forces. Everyone felt pretty good.

A black-and-white road sign went past our Humvee. On it was the name STE. URSULE DU LAC.

Only an optimist would have called it a village. It was just half a dozen houses and a school grouped around the Norman church of Saint Ursula. It was deserted.

The Union had something they called "police battalions." They came in behind the fighting units, and when an area had been pacified, they were supposed to stay behind and make sure that law and order were restored.

That happened, sometimes. More often, the police battalions were just a euphemistic way of solving the knotty problem of what to do with an Occupied and presumably annoyed civilian population. As the Union advance pressed westward to the Atlantic, they had left hundreds of empty villages in their wake. We'd been on the trail of this one particular battalion for a couple of days now.

Nobody was under any illusion that the Alliance forces were any better than the Union; there had been atrocity on both sides. But as journalists, we knew which side our bread was buttered on. We were traveling with the Alliance; we were hardly going to file stories accusing them of human rights violations.

We pulled up in Saint Ursula's little village square and dismounted from our various vehicles, stretching our legs. The Alliance had already come through here a few hours earlier and pronounced the coast clear, but soldiers fanned out to search the buildings while we journalists stood around smoking and chatting and doing pieces to camera. Someone unpacked a portable catalytic stove and brewed coffee. The smell drifted on the breeze.

I wandered away from the main group. None of the buildings in the village seemed to be damaged. There was no sign that the War had come this way at all. But there were no villagers. There wasn't even a stray dog.

The school was a little way up the single street from the square. I lit a cigarette and put my hands in my pockets and walked up to it. It was white, and there was a little black bell mounted on a swivel over the front door. I walked up the steps. Someone behind me was shouting.

I looked over my shoulder. One of the Alliance officers was running toward the school, shouting something and waving his arms. It was the little ginger-haired major from New Brunswick, the one who claimed he'd worked on the *Chicago Tribune* before the War. We all thought he was a dickhead and did our best to ignore him.

I turned back to the door, turned the handle, and pushed.

"Anyway," said the ugly little man on the other side of the desk, "it's not very much, but it's something." He smiled awkwardly. "It'll keep you off the streets."

I looked around me and blinked hard. I said, "Did you just offer me a job?"

Somewhere, between pushing open the door of Saint Ursula's school and waking up in Rex's office, three years had passed. I didn't know how I had returned from France. I didn't know the War was over. I didn't know the elves had taken control of Britain.

I had turned up in the village a week or so earlier, "an animal dressed in rags," as Liam put it. The Village Council didn't know quite what to do with this raving madman. They'd cleaned me up and fed me and, when I didn't seem too dangerous, Rex offered to give me a job at the *Globe*'s offices, sweeping up and moving rubbish and stuff.

I didn't know why I came out of it when I did. Maybe Rex said something that brought me back from wherever I had gone to hide.

I didn't know what I saw when that school door swung open, but late at night, when I was lying in bed, terrible things beat on the thin walls of sleep, looking for me.

I opened my eyes.

There was a smell of burning in my bedroom.

I sat up. The light of a full moon was flooding in through the windows and falling on an elf, which was sitting on the end of my bed, smoking a spliff.

I shouted something and flopped back onto my pillows.

"You're looking well," said the elf. "Newspaper work obviously agrees with you."

I said, "Did anyone see you come in?" Elves were not the most popular people in Britain. If anyone had seen this one enter my house, the most optimistic thing I could look forward to would be a vigorous lynching.

It took a huge toke on the joint and blew out a stream of smoke that was silver in the moonlight. "It's half past three in the morning," it said. "Anyone out at this time of the morning isn't going to believe they saw me, even if they did. Which they didn't."

I sat up again and mashed the pillows down behind my back. The elf called itself 56K Modem. That wasn't its real name, of course. The elves took whatever pleased them, including their names. Modem once told me its real name. It sounded like snow settling on a frosty road.

"What do you want?"

Modem tapped ash onto the floor. "Aren't you going to offer me a drink?"

"Do you want one?"

"No. But I'm rather hurt you didn't offer."

Modem was wearing a collarless white shirt and jeans. Its feet were bare, and its fine gray hair was bound into a meter-long rope. I rubbed my face to try to wake myself up. "What do you want?"

"I heard that Rex killed the pig."

"Everyone else knows about it. Why shouldn't you?"

"It's an interesting situation, don't you think?"

"It's a fascinating situation, but I really need to get some sleep, so I'd appreciate it if you'd come to the point." Modem had been visiting me, on and off, for a couple of years now. The first time, I had tried to run away screaming, but these days I was almost blasé about it, as if I weren't sitting in the same room with one of the most dangerous predators on the planet. I could even do small talk with it.

On the other hand, I had never found out why Modem visited. It usually spent its time taking the piss out of us, telling me how pathetic we were and how brilliant the elves were. I had a feeling—and it was nothing more than a feeling—that, somewhere in the black hole of memory between Saint Ursula and Belton, I had done something for the elves, or been forced to do something for them. It was a prospect that brought me out in a cold sweat.

Modem looked at me and tipped its head to one side. The moonlight made it look ethereally beautiful. "We were wondering if you'd like us to intervene."

"No. Can I get some sleep now?"

Modem looked hurt. "Are you sure?"

"Yes."

The elves hated Mankind, of course. They had been masters of the world for uncountable centuries. And then we had come along with our *technologies,* and we had cut down the forests that were their natural habitats, driving them back and back until all they could do was watch as our cities were built and wait until we were vulnerable.

"We're quite interested in Rex Preston," Modem said.

I felt ice touch my heart. "Oh?"

The elf uncrossed its legs, recrossed them, brushed a piece of lint off the thigh of its jeans. "Actually, I'm interested in what you think of Rex Preston."

I looked at it. "Why?"

Modem thought about it for a moment. "Professional curiosity?"

"He's a newspaper editor. How on earth can you be professionally curious about a newspaper editor?" The first rule about the elves was this: You didn't annoy the elves. That was the only rule, really, but I'd learned that there was some latitude. You could annoy some of them more than others; it was just impossible to tell which ones. You had to wing it.

"His paper is on its knees. His wife works for his competitor. He just killed his last animal. But he won't give up." Modem tipped its head to one side. "Personally, I find that kind of . . . devotion interesting. I've noticed something similar in the Resistance."

I burst out laughing. The Resistance was a largely theoretical thing, armed with whatever weapons they could scrounge from the days when the Alliance was based in southern England. They killed elves here and there—on the orders, legend said, of an ex–New Zealand Special Forces Colonel who had found himself stuck here just after the War. For every Resistance success, the elves destroyed a village or a town. Popular opinion had it that the Resistance had caused more loss of life than the elves themselves.

"Rex isn't with the Resistance," I said. "It would get in the way of putting the paper out."

56K Modem looked at me and pursed its lips. "All the same," it said, "perhaps he would bear watching."

The elves had something roughly analogous to MI5. They called it The Library, and among other things it was charged with dealing with the Resistance. They hunted down ham radio operators and ham TV operators, they hunted down people who put together kit-cars in their garages or played guitars and sang to each other late at night. I thought it must be a pretty thankless task, but Modem seemed to find it fulfilling.

"Rex isn't with the Resistance," I said again.

"Harry Burns is."

More ice around my heart. "Harry's not with the Resistance either."

"He's at a meeting right now," Modem told me. "Over on the outskirts of Sheffield. There are five of them in a house in Dore. They're planning an assassination. We disapprove of assassinations."

I shook my head. "Not Harry."

"Harry's ex-SAS. Good with munitions." Modem blew gently on the burning coal at the end of the spliff. "An absolute star. No end to the things Harry can do with a few ounces of plastic explosive."

"What do you *want*?" I shouted.

Modem looked taken aback. "It's just this situation with Rex and Liam—"

"Yes!" I yelled. "Intervene! Do whatever you fucking well *want!*" We sat looking at each other from either end of the bed. "Are you happy now?"

Modem stood up. "I'm never what you'd call *happy,*" it told me.

Every village has a character. Sometimes, if the village is big enough or unfortunate enough, it might have more than one. Ernie Hazlewright was ours, a big, permanently annoyed old man who lived just down the road from me. He was a legendary drinker and a brawler of some note, and he'd been barred from all three of the pubs in the area more times than anyone could remember.

By rights, he should have gone down fighting in a punch-up in the street, but he'd actually fallen into the river while walking home pissed out of his mind one night and drowned. I supposed it was a rather sad way for a Falklands veteran to go, but I wasn't going to miss him.

Still, it was rather a good turnout in the little cemetary down by the river. About thirty people turned up, mostly Ernie's old drinking buddies. I managed to get a few words from each of them.

The mourners had all gone off to the pub, and I was chatting to the vicar when I saw Rex coming down the gravel path from the church. He stopped by Ernie's grave and stood looking down at the coffin. I went over to him.

"He wasn't a bad old lad, really," he said. "Just drank too much."

"He was an absolute nightmare," I said. "Coming home legless at all hours of the day and night, beating up his wife. You didn't have to live near him."

Rex nodded. "That's true."

"He smashed all my front windows once."

"You haven't been in to the office yet, have you?" he said.

I shook my head. "I came straight here."

"So you won't have seen what we found in the yard when we came to work this morning."

"No, of course not."

"So the animals didn't have anything to do with you, then."

I frowned and felt my stomach start to contract. "What animals?"

Rex shrugged. "Well, I left Harry counting them, but it looked like fifteen or so chickens, half a dozen goats and four pigs. Three sows and a boar."

I stared at him.

"Anything to do with your *source,* do you think?"

I had never kept anything from Rex. I had told him everything about myself, at least everything I could remember. He was the only person I had told about 56K Modem and its visits, and I thought it was probably the bravest thing I had ever done. Rex, of course, was an old-fashioned sort of newspaperman. A contact with the elves was literally beyond price, even if it might be morally suspect, and a good journalist always protects his sources.

"Modem came to see me last night," I said. "It asked me if I wanted them to do something about this thing with you and Liam."

Rex frowned. "Why?"

"I don't know. It's a game with them, Rex. They think we're *funny*. They watch us like we're some kind of soap opera or something."

He scratched his head. "Well." He turned and started to walk up the gravel path toward the entrance to the churchyard. I followed. "I'm not sure whether to be flattered or not."

"Best not."

"Aye, maybe you're right."

"Someone had better mention to Harry that the elves are on to him, as well," I said.

He glanced at me. "Did your *source* tell you that, as well?" I nodded, and he shook his head. "Why don't they just pick him up, then?"

"I told you, they love to play games. Modem said Harry was in Sheffield last night meeting with a Resistance cell."

Rex put back his head and laughed. "Either your *source* was playing games, or it's not so well informed as we thought. Harry was nowhere near Sheffield last night."

"Oh?" I was rather hurt. "Where was he, then?"

"He was with me, burning down the *Chronicle*'s office."

I stood still. Nobody I had spoken to this morning had mentioned anything about a fire, but I supposed they'd all had other things on their minds, like mourning Ernie and getting to the pub for opening time.

Rex walked a few more steps; then he turned and looked at me. "Don't just stand there with your mouth hanging open," he said. "*He'd* have done it to *us*."

That was fair comment, I supposed. "You'll never get away with it. He'll know who it was."

He smiled cheerily. "There's *knowing*," he said, "and there's *proving*."

I caught up with him. "I didn't think you had it in you."

Rex looked thoughtful. "No," he said finally. "No, neither did I." He looked about him, then started walking again. "We didn't put him out of action permanently, anyway. Just sort of charred the office a bit. He'll be up and running again in a couple of weeks. It's actually rather funny."

"How on *earth* is trying to burn down your competitor's office *funny*, Rex?"

He chuckled. "It's just that Harry and I spend most of the night skulking around the *Chronicle*'s office, trying to put Liam out of action for a while so some of the advertisers will sign up with us, and this morning we find the backyard full of livestock, courtesy of your friends." He shook his head. "I just found it funny, that's all."

I wasn't so sure about that. I knew there would be some kind of price to pay, but there was no way of telling what it would be. Or what would happen when the elves didn't find us amusing any longer.

"One of us ought to go over to the *Chronicle*'s office and do some kind of story," Rex said.

I grinned. "You cheeky old sod."

"It'd look suspicious if we didn't. And it's good copy, anyway. 'Local

Newspaper Burns Down.' " He nodded to himself. "Good local copy. The cornerstone of a good local paper." He looked at me. "Would you like to do that?"

"It would make my day," I told him.

He nodded again. "Good lad. And if you see Alice, give her my regards." And he walked away, head up, back straight, whistling a little tune, the happiest editor in Derbyshire.

John Meaney is the author of three groundbreaking British novels. *To Hold Infinity* was one of the *Daily Telegraph*'s Books of the Year. It was also nominated for the BSFA Award, as was his acclaimed second novel, *Paradox,* which begins the Nulapeiron sequence, continuing in the sequel, *Context,* and the forthcoming *Resolution.* Meaney holds a degree in physics and computer science, a black belt in shotokan karate, and is addicted to weightlifting, running, languages and science, all genres of fiction, cats, coffee, and chocolate.

THE SWASTIKA BOMB

John Meaney

A great black delta-shape slid overhead, gut-wrenching subharmonics pouring down in waves. It slewed into position, hovering above Nelson's Column. Panicked, pigeon flocks exploded outward, a burst of wing flurries as the intruder's shadow fell upon Trafalgar Square.

Heinkel Drache 22-E. I recognized the species immediately. Already, beneath its wings, the dark, deadly payload was struggling to be free.

I stared upward—frozen, despite all my training: the wings' noise felt *solid,* beating down—and swallowed at the sight. Great Luftwaffe cross-insignias matched the black wriggling *Hakenkreuzen* slung in bomb racks.

Move.

Wrenching my attention down, preparing to run . . . and right at that moment I saw her, and fell in love.

Like a spring storm, when the sky darkens, yet suddenly a figure on the ground glows in contrast with white, internal radiance—that was her. Upswept honey-hued hair, her pale triangular face an ivory glow above the trim blue-gray RAF uniform, wide-eyed . . .

Her devastating gaze met mine.

My God. Such beauty—

Then the Swastika Bombs dropped.

She hesitated beside the dry wide fountain—the black lion statues were at my back—as I *felt* the movement overhead, and then I was moving very fast.

By the time I reached her, she had kicked off her shoes, and I grabbed her jacket in the small of the back—that perfect back—and lifted her almost bodily as we sprinted in time, swerved, ran hard, heading for the tall pale church of Saint Martin-in-the-Fields as the big black hooked crosses fell.

They snapped, writhed, anxious for the kill.

Run.

Her thoughts were mine, in total communion, with no breath to speak: we were elemental primates, running for the mortal joy and fear. Sprinting for our lives.

Faster.

Then we were at the building's side, and for a moment I thought we might run on to Charing Cross, but she tugged left, making the decision, and we skidded to a halt beside piled sandbags as a great *crump* sounded from behind. Ravening bombs, like great conjoined black worms, whipped in their death throes, smashed granite like papier-mâché, building up a crescendo of maddened destruction, internal pressures building, rising until virus-venom sacs and acid bladders exploded—

Too late.

But we were already stumbling down the steps, and helping hands dragged us through the doorway and then we were inside the church's darkened crypt and safe. A great gout of dust followed us in as walls blew apart on the street above.

Safe . . .

And in that moment of exhilaration, we grasped each other hard, and kissed—her lips like silk, absorbing—each pulling the other inward as though we would fuse together in blissful joy, forever.

For this was war, our darkest hour, and things were very different then.

I asked her, "Will you marry me?" and she said, "Yes," and then we told each other our names.

This was the way of things, when no one knew whether life or death awaited the next day, or the next.

"I'm Laura." Her voice was breathless and elegant.

Around us, others crouched or stood in the shadow-shrouded crypt, barely lit by flickering orange candle flames. Some were civilians in mufti; most were service personnel in uniform. They hunched atop the three-century-old headstones that paved the floor, sunk inside their own thoughts or muttering to those close by.

But to me, there was only Laura: the real and beautiful wonder of her.

"Listen."

Her hands tightened their grip, and then I heard it: a bright silver whistling through the air.

There was a grim edge to her smile as she whispered: "Now we'll see."

A distant *whump,* and you could picture the sudden burst of flame, feel the wave of heat. Then we all heard the strange eldritch cry of a wounded bomber-dragon.

"Spitfires!"

Cheers rose up around the darkened crypt.

"Bastard Jerry's had it now."

Later, we would get the full description: of those brave young men who took their green-and-silver raptors, perched in the tiny spinal cockpits, melded into their mounts' ganglionic pathways, hurtling down from the clouds with wings furled, then snapping those wings outward—under heart-wrenching stresses—as they swooped past the formation of great black delta-winged dragons, glimpsing the crews' shocked faces as they carried destruction back to the death-bringers.

It was not one-sided, that fierce battle, for the Luftwaffe had their own escort of emerald-green Messerschmidt Falke-104s, the dreaded Falcons glinting in the sun, and soon there were duels in the air above beleaguered London as the faster Germans took on maneuverable Spitfires: brave young men—on both sides—with iron will and lightning reflexes, fighting for their countries, risking everything.

The bomber-dragon that had hit Trafalgar Square took severe injuries in those first few seconds: from flamebursts and fling-stings as the Spitfires swooped past. Wounded, the great black dragon swung up into the air, delta wings curling, but the injuries were mortal and the crew could do no more as vital organs burst inside, and then they were falling in a long lazy arc towards the Thames.

It smashed into waves, the impact hurling gouts of spume high into the air, wrecking any hopes of capturing the aircrew alive: pulping their bodies as the dragon's back broke in two.

Later that evening, crowds of adults, and a small number of children who had missed evacuation, would watch from the Embankment as navy trawlobsters dragged the huge corpse ashore, to the waiting RAF dissectors.

But as the air battle, still raging overhead, moved farther away, there was a ripple of relieved laughter in the crypt, the lighting of cigarettes, which had been ignored until now. The air seemed easier to breathe.

"Here you are." Laura handed me my spectacles. "If you want them."

The glasses had survived the run, only to fall upon a worn seventeenth-century gravestone beneath our feet. One lens was splintered.

"Thank you." I tucked them inside my suit jacket—double-breasted, brown, now stained with dust—and looked around for my hat. "I might need them later."

"For disguise? This close to Whitehall?" Those elegant lips curled into a smile. "Or have plain-glass lenses got some use I don't know about?"

"You're very observant."

"Not only that . . . I suspect we're headed for the same place, my love."

Then the all-clear sirens wailed, and it was time to go.

As we passed through the square, the skies were clear save for a distant observation blimp, blue gills fluttering as it scoured enemy spyseeds and germspores from the air.

Hand in hand, Laura and I skirted still-moving sections of black rubbery Swastika Bomb carcasses, while bomb-disposal scarabs crawled around the debris, directed by their asbestos-suited masters.

Cleanup dromedatanker crews were already spraying antiviral mist across the foul yellow fluids spattered everywhere.

We walked carefully, avoiding acid puddles and toxic pools, and turned into Whitehall. The long boulevard was remarkably untouched by the action. Tall white-gray buildings stood blocky and proud; their cross-taped Regency windows, with their heavy blackout curtains and sandbag reinforcements, concealed the smoke-filled rooms of the War Ministry, the Admiralty spillover, and Military Intelligence.

They formed the intellectual center of the British war effort, in those darkest die-before-surrender days when it seemed certain the Blitzkrieg would destroy us all.

At a short flight of pale steps before tall iron-studded doors—we *were* headed for the same place—Laura kissed me lightly on the cheek. Then we passed inside, Laura returning the sentries' salutes, while I nodded politely, reinforcing my pseudocivilian status.

In the foyer, as my heart shivered, we parted company for the first time.

Blue pipe smoke wreathed the Old Man's office: a dark, mahogany-paneled chamber, heavy with stern predecessors' portraits. From the anteroom, I peered in, and waited while a group of women, half in uniform—WAFs and FANYs—exited in a clatter of high heels, a businesslike bustle of long skirts, clutching their clipboards.

All intelligence personnel had antiallergen infusions and oncovaccination. Still, there were reddened eyes; everyone looked relieved to be out of the smoke.

"Come in, old chap." The admiral's voice boomed from behind his huge desk. "Take a pew."

"Sir."

I nodded automatically to his secretary, the formidable Miss Poundstone—chestnut brown eyes, so different from Laura—and passed inside, closing the heavy doors behind me. He indicated a chair with his pipe stem.

Laura . . .

"Now then, Fleming—"

The Old Man's words snapped me back into the moment. "It sounded urgent, sir."

In the hard-backed chair, I waited for the fearful words.

"Time"—his big blunt fingers drummed on the leather-bound blotter—"that you went back into the field, don't you think?"

No. Too soon.

Part of me had had enough: of the twilight-sleep treatment in a Birmingham hospice on my return, drugged to the gills again, with morphine and God-knows-what; then of the endless retraining in the windy Rutland countryside, restringing my nerves in the echoing former public school where operatives underwent the harsh discipline designed to keep them alive. I had been teaching, too: as an old hand who had survived—flea-bitten and ragged-eared, but functioning—giving the cynical benefits of my hard experience to wide-eyed neophytes who revered me too much.

For their eyes were bright with youth's secret knowledge of their own immortality, secure in their ironclad conviction that the torture victim's screams as Gestapo interrogators bent to work, the broken corpse splayed facedown in a dark puddle in some cobbled alleyway, were images from someone else's future, not from theirs.

"You're sending me into occupied territory?"

Part of me yearned to return: that desperate obsession with the brink. Yet all of me wanted Laura: my world was different now.

"Where else?" With a frown, "We can only mount this operation with a particular type of agent at the forefront. That means you, Fleming."

"Thank you, sir." I took it as a compliment.

"You know about nanoviral vectors. Well"—from under heavy white eyebrows, he glowered—"our cousins across the water are putting their best boffins—and I include the best who escaped Europe, not just Yanks—onto a single weapons program. The Brooklyn Project promises to end the war at a stroke."

I shifted. A feeling of dread slid across my skin.

"You don't want to hear this." For all his bluff manner, the Old Man could be sensitive to the point almost of telepathy. "Strategic context is hardly de rigueur in briefings."

Worrying about long-term implications could make your trigger finger tremble at the wrong moment. Too much background was dangerous.

"The thing is"—with a granite smile—"you'll pick up the resonances soon enough, so I'm telling you right now."

His sympathy worried me.

And there's Laura to think of now.

But she was no more a civilian than I was. The war was bigger than me, or Laura, or any of us.

"What's wrong with this project?" I asked the Old Man then.

"Nothing, per se. But the Nazis have their own program running. If they beat us to it, the world will be under Adolf's jackboot before you know it. And we're not going to let that happen."

"No, sir."

Those two defiant words were my commitment: to the mission, to the future, to the decisions that changed our lives forever.

I had a pass that would let me through any Allied checkpoint, even if the name read H. Himmler and I was belting out the "Horst Wessel" song at full volume. That was the one I used now, accepting the sergeant-major's salute.

I waited while the megarhino-hide blast doors slid open, then descended hard chitin steps into the Tac Bunker.

The observation balcony already held a visitor. I recognized Admiral Quinn; it was not mutual. Instead, he nodded cautiously: a typical senior officer's greeting to a Whitehall man dressed in civvies.

He slid a cigarette half out of his packet of Senior Service, and proffered it.

"No thank you, sir."

"Fascinating." He closed the packet without taking one himself, slid it inside his dark braid-decorated jacket. "Don't you think?"

"Always."

Beneath us, the cavern stretched for two shadow-shrouded miles, upheld by massive columns, spanned everywhere by darkly glistening fibers that formed the Black Web. At the lowest levels, Strategic Command personnel were tiny uniformed figures scurrying amid overwhelming complexity, clutching their clipboards like talismans. Some walked with blue-scarlet messenger-transparrots perched upon their shoulders, ready to fly upon command.

But it was the Web itself that drew one's gaze.

Black hawsers as thick as Nelson's Column, a plethora of others merely as wide as my forearm, down to millions of near-invisible threads: all dark as night, catenary curves hanging in a three-dimensional maze, beyond any single person's comprehension.

And along those threads crawled, in every conceivable direction, the one army that might halt Hitler's destruction of Europe: dark micro-spiders, and their bigger, fist-size counterparts whose swollen thoraces would burst open to give birth to myriad offspring. Each tiny arachnid speck would already be imprinted with interaction behavior; combined in their millions, the combination formed the Black Web's ghostly, disembodied gestalt: a paramind that knew nothing of fatigue or fear or morality, but whose analytical powers might someday grant us victory.

"Excuse me, Admiral."

My footsteps sounded strangely hollow as I passed along the chitinous catwalk, descending to the North Atlantic display. A wide map table tracked the vital Liberty Leviathan convoys, the U.S. lifeline that—bless their hearts!—was all that kept the British Isles from starvation.

But we had only patchy success in tracking the enemy's U-shark hunter-killer squadrons, which preyed on merchant vessels from beneath the gray choppy waves, too often taking out the battleorca escorts that formed the backbone of His Majesty's Royal Navy.

And I thought of the Yanks' great behemoth-class carriers, burning and squealing in their death throes, trapped helplessly in the docks of Pearl Harbor beneath the Japanese hydra attack, the ravening dragonfire of Yamamoto's carrier-launched squadrons. We weren't the only ones to suffer in this war.

Military considerations fled my mind as I saw that elegant figure bending across the table, biting her lip as she used a long pointer to push a tiny model convoy onto revised coordinates.

Laura . . .

Her name was a prayer, a sigh.

When she looked up, a smile broke across her face, and she waved.

* * *

The tiny tearoom was spick and span—no ring-shaped stains, no chipped cups—with whitewashed bricks. Close heat emanated from large steam pipes running beneath the ceiling, occasionally knocking as if someone were trying to get out. The kettle's long flex ran across hooks above a poster reading THE ENEMY HAS SPIDERS EVERYWHERE. Overstuffed armchairs and a tattered settee lent it the look of Senior Common Room, from the days of schoolboy innocence.

Before '38 . . .

Laura touched my arm. "Are you all right?"

"Just a touch of malaria." The usual excuse—but this was Laura. "Actually . . . A touch of memory. That's a whole lot worse."

November: a chill, fresh breeze blowing along Friedrichstrasse—

Even now, it haunted me.

—with morning sun glinting on the crystal shards.

"The Old Man's sending you back out."

I looked at her. "Just what did you say your assignment is, darling?"

"My current job involves technical background for someone who's on a trip to the American Southwest."

"Does it, indeed?"

I was destined to meet you, my love.

"Oh, yes." She handed me a cup, balanced on its ugly flower-decorated saucer. "Have you heard of Albert Einstein?"

"Well, yes. He's famous for—"

"The Project is his idea."

"I'm sorry?"

"He persuaded Roosevelt. Signed an open letter, with several of his colleagues from Princeton and elsewhere. The Brooklyn Project grew directly from that letter."

"I didn't know."

"Einstein's a true genius. Special multiplicity theory is what everyone knows him for—"

"DNA structure," I said. "Replicator-based evolution. Gene migration."

Laura, sitting with her knees primly together and her tea balanced atop, nodded as though everyone knew that. Above us, the steam pipes clanged.

"The point is—" She glanced up, waited for the noise to subside. "—the special theory was inevitable. Yet *general* multiplicity theory is a stroke of genius, ahead of its time."

"If he hadn't been born . . . Or if he hadn't been Jewish, he might've been a Nazi."

"Don't say things like that." Laura shivered.

The tea in my mug had grown cold, and I set it down on the parquet floor. "Sorry."

She laid her hand on mine. "It was luck that you and I met now. We weren't supposed to . . ."

My world would have remained bleak and Laura-less, forever empty.

"Perhaps SOE didn't plan it, but Providence did."

Outside, footsteps sounded. We waited, but they walked on past the doorway.

Laura touched my cheek.

I leaned forward, and we kissed.

One week later, I was standing at the edge of a pink-orange desert, beneath a cloudless azure sky, a white-hot blazing sun. The air's furnace heat pressed upon me, squeezed me in its invisible fist.

A dark road arrowed across sands, disappeared over a low ridge. Behind me, straight railway lines—rail*road*, I should remember where I was—gleamed silver through the rippling air. The diminishing train slid through heat-haze, its chugging oddly flat in this dryness. I watched it disappear.

And then I was alone in a sere, endless landscape where human beings were never meant to live.

At my feet stood a battered suitcase that had been with me in the Far East and occupied Europe, though it bore no labels to mark its travels. My woolen suit, perfect for an English spring, was heavy upon me; folded atop my suitcase, the big overcoat looked ridiculous.

As was this rendezvous: away from the town, and the eyes of hypothetical fifth columnists. I could hardly be more conspicuous . . . should there be anyone here to observe me, before heat and dehydration took me.

I loosened my tie.

Off to one side, purple mountains shaded the horizon. Would there be snow upon their peaks? The thought of frozen water melting in cupped hands was almost overwhelming. Much closer, hundreds of saguaro cacti stood with upraised arms, as though caught in a holdup: like green capital *psis*.

Braithwaite's ruler snapped me across the knuckles as I stumbled conjugating my Greek verbs—

Those damned school days. They haunt you forever.

Stop whingeing.

In Africa, thousands of men were facing worse conditions than these. At least here, there was no threat of Rommel's rextanks suddenly appearing over the horizon, bringing their heavy armament to—

There.

A small dust cloud puffed upon the distant blacktop.

Deliberately unarmed, I waited in the heat. If they were an enemy of some kind, they would not know my face; at close quarters, the absence of a weapon might convince them I was harmless. And I always had my hands.

Half-fist to the gendarme's larynx. He falls, croaking, clutching his throat, eyes popping as he chokes. . . .

A bad memory. But it had been a necessary killing, and I got away before the Waffen SS officers arrived: the ones I was sure—*almost* sure, that's what tortured me—the gendarme had notified. In the darkened alleyway, with the rain falling down in sheets—

Concentrate.

Here. Now.

Olive-colored, the jeepo was much closer now, revealing the driver: white shirt, broad scarlet tie blowing in the slipstream. I swallowed, wishing I could be that cool.

And then it was skidding to a halt, the jeepoceros bracing its squat, powerful legs, throwing up a cloud of dusty orange sand. Sunlight glinted—*watch out*—but it was not a weapon: the driver bore a polished steel hook in lieu of a left hand.

"Climb in, pal."

I controlled my breathing, ready for action. "It gets cold," I said casually, "in December."

"Yeah, right. Willya get in?"

I waited.

"All right . . . Drops to seven below." Hawklike features frowned. "Or is that nine? Like I should know Latin, too."

Parole-and-countersign were a random numerical reference, indirect if possible—I granted that *December* might be ambiguous—and a reply code which subtracted three. Where I came from, we took such things seriously.

Throwing my case and coat into the back, I slipped into the passenger seat (on the right) and held out my hand. "How do you do."

"Hi. The name's Felix. Felix Leichtner, though I'm thinking of changing the name. My old man will be pissed."

"I would think so. I'm—"

"I know who you are. Hang on."

He threw the switch, and the jeepo rumbled into life; then he swung it through an impossible turn—I glimpsed a small tan form watching us: ground squirrel—and floored the accelerator. Dashboard membranes flared red as we tore off along the lonely road, slipstream blowing, dust cloud billowing in our wake, while in every direction the orange sands stretched hot and majestic: a wilderness that could kill as surely as a hail of Wehrmacht bullets.

Vastness.

Overhead, a lone eagle wheeled in the deep azure sky. If the jeepo failed, would vultures make their own ghastly appearance? On every side, sandstone glowed; farther back, rocky outcrops bared strata: mint-green and sugary white, stippled with black.

Would Laura care to live here after the war, with me? Picture it: an adobe house near a small desert town, where I could paint New Mexico landscapes for a living, and share this beauty with a woman beyond compare.

In the midst of vastness, an outpost.

We shot under the candy-striped barrier before it fully lifted—Felix taking time to give a shoddy half-salute to the sentries, his hook-hand steering—and onto Main Street: the largest straight dusty track between wooden rows of whitewashed houses. Their plain exteriors held the charm of military barracks everywhere.

But soldiers in camp did not have individual homes, with slender wives in light cotton dresses—

"Eyes on the road, Limey."

"You're driving."

"But that's Mrs. Teller, and you don't want to start off on the wrong foot, my friend."

I shook my head. "I'm spoken for."

"Not what I heard." Felix glanced in my direction, then nodded as he span the jeepo to a screeching halt. "Good for you, though. And this"—with a blinding glint from his gesturing hook—"is home, sweet home."

Two badly upholstered armchairs sat in the too-hot lounge. We dragged them to face each other—I had unpacked bug spray and used it: no eavesdrop-mites would survive the aerosol—so that Felix could brief me.

"First off, the Oppenheimers are throwing one of their parties tonight. You'll like Oppy. The Martians will be there—"

"I'm sorry?" This briefing had suddenly veered onto an unexpected track.

"Pet name. A bunch of Hungarian scientists: Wigner, Szilard, Teller, von Kármán. They're hoping to get Johnny von Neumann on board, from Princeton."

"I don't know him."

"Fun guy, sort of. Wife called Klara, a bit shrewish. He's something like your Turing, in his research interests. I think they met briefly."

I shrugged, but—inside—my nerves strung themselves tighter. *Plying me for info?*

"Come on, pal." Felix sighed. "Turing's geneticists at Bletchley Park are your major contribution to the war effort, and you've got the right level of clearance to know it. Without him, the Nazi commspiders' genomes and ganglionics would still be a mystery."

I relaxed, a little. "It was *your* clearance I was worried about, old fruit."

Felix gave me a strange look.

"That's not an insult, is it?"

"God, no. Look, can we go somewhere to get a coffee?"

"Sure." And, as we stood, "I thought you Limeys only drank tea, old chap. Sorry . . . old *fruit*."

I shook my head, but in fact I wished that Felix was not going to leave me here, among the boffins and their wives. I would rather spend my days swapping Yank-and-Limey humor than bashing my head against cutting-edge military research.

Laura . . . It's your company I need.

But if we survived this ongoing nightmare, formed the future we both wanted, it would be worth the wait.

Secrets within secrets.

Even in the long mess hall, we talked only of innocuous matters, not trusting the background chattering and clattering to hide our words. Felix noticed my interest when a platoon ran past outside, to cadence.

They were lean and fit to a man, and I needed to keep my own physical levels high. A tan would be suspect, once they dropped me into Europe; but melanin-reversers could wind back exposure effects from the desert sun.

After we returned to my temporary home, Felix briefed me on base security. "The core," he said, sitting down, "is a team of what are supposed to be Airborne Rangers."

"But—?"

"They're some of Donovan's finest, and they keep closer track of both people and research than the intellectuals expect."

"I may not be intelligent"—I was quoting the Royal Navy gunners—"but I can lift heavy weights."

"Like that, yes. Easily underestimated."

So operatives from OSS were here. Very interesting.

"Can I train with them? I need to keep fit."

Dawn runs around Hyde Park, evening Indian club practice: they kept me going in London. But I needed more, and the Rutland training school was half a world away.

"If they can find you a uniform . . . They get up real early, you know?"

"Good. I'll be busy working my brain the rest of the time."

"You don't know the half of it. Dick Feynman's a magician, literally and intellectually. He'll mentor you for an hour or two a day, but you'll be studying like crazy on your own."

"Good . . . What's he like?"

"Who, Dick? Quite the prankster. Cracked open the base commander's safe, left a birthday card, with a French letter tucked inside."

I rubbed my forehead with the heel of my palm. Travel-lag, after the long crossing in the deHavilland pteradrone, and the long train journey.

"This is supposed to be serious."

"Feynman's wife is in a sanatorium, near the town. Dying, apparently. The comedian act is partly a front."

I blinked and turned away. *Laura . . . I miss you.*

"Resistance-engineered bacterium. He hates the Nazis more than anyone I've met."

"All right." *But was he there for Kristallnacht?*

I said nothing, but my guard was down and Felix sensed the vibrations.

"Though you might give Dick"—his hawk-smile was humorless—"a run for his money."

"Chattanooga Choo Choo" was blaring from the radiogram as we insinuated ourselves into the party. From their files—I'd been granted access; Felix stood over me in the base commander's office as I read—I recognized faces: Oppenheimer, Fuchs, Teller.

Off to one side, lean and quick-witted, Feynman was dazzling a group of his colleagues' wives with a conjuring trick. His sparkling eyes, more than sleight of hand, captivated them.

"Not a security risk, then?" I murmured.

"Probably not." Felix was looking at the buffet.

I knew what he meant. There was no reason for a fifth columnist to make himself conspicuous in quite that way: no reason to perpetrate double-bluff. But schizotypal behavior was common among academics. . . .

Natural paranoia: discount.

A pretty woman offered us drinks. Her brilliant smile, directed at

Felix's lean face, became fixed as he tapped the glass with his steel hook.

"Nice crystalware."

"Um, thanks . . ." And then she was gone.

I looked at Felix. "That was a little coldhearted."

"Gets things out in the open." He gave the tiniest of shrugs. "It doesn't worry everyone."

"What do you tell them?"

"Just a wound I picked up in—Oh. Massive allergies, is all."

I nodded. OSS personnel were so thoroughly infused with antivirals that their bodies threw off clone-factor treatment.

"Why, *Fred*—" There was another smiling woman in our path, looking impossible to faze. She used Felix's cover name with irony, as though aware it was not his real identity.

"How're you doing?" said Felix.

"Why don't you introduce me to your friend?"

Courtliness, not flirting.

"The name's Brand." I used my own cover ID. "James Br—"

"And he's a Limey," interjected Felix. "But he can't help it."

"Oh, that's wonderful."

"Jamie, this is Mrs. Oppenheimer."

She held out her hand to shake.

For a while we exchanged pleasantries as she pointed out people of interest—"That's my husband, known as Oppy to these young people"—and made rueful observations about the poor facilities in their makeshift town.

"But we're coping," she said. "And it's important work, we know."

Then someone caught her attention, and she bade us a polite farewell and moved on. Felix and I remained standing, while a few of the more energetic men danced with their wives to Glenn Miller. At home, they'd have been singing by now, about hanging out their washing on the Siegfried Line.

It was too hot to party. I was relieved when Felix took my elbow and steered me toward the screen door, passing close to Feynman and his admirers.

"—be kidding, Dick," one of the pretty women was saying. "That's disgusting."

"Nothing personal, ladies." Feynman grinned. "*Everybody's* body weight is ten percent bacteria. Not just yours. They're wriggling about, swapping genetic material like crazy—"

Shocked intakes of breath. Giggles.

"You look at the world, Dick"—the woman who spoke was pale-complexioned; she briefly touched Feynman's knee—"like nobody else."

(Was I the only one who saw the shadow pass behind his eyes, the thought of his dying wife?)

"But there's beauty in that dance of life, don't you see? You can understand the way things are built, and *still* enjoy the way they look esthetically."

"Uh-uh. You're just different from the rest of us."

"I'll say." One of the other women smiled.

Everybody laughed.

"Except that"—Feynman was excited now—"we can't be all that different from each other, or you couldn't build a human being by taking chemical fragments of two people—*unrelated* people, unless there's anyone here from Kentucky?—and fusing them together. That's exactly how babies are made."

"But Dick, I thought—"

And then we were past them, through the door, into the desert night.

Overhead, black velvet and a blazing multitude of silver stars such as one could never see in England. From behind us, a burst of laughter.

"Makes you realize," murmured Felix, "how insignificant we are. Maybe humanity deserves to go under."

I glanced back into the warm light, the partying scientists. "And is our salvation really in there?"

"Getting drunk and flirting? Why not?"

We stepped into the middle of the pale moonlit dusty track that passed for a road.

"Y'know what?" Felix added. "A couple of them have calculated that, if the nucleic bomb goes off, the reaction could cascade through the biosphere. Wipe out all life in hours. And I mean *all* life."

I stopped. "You're jok— No. Is Feynman one of them? One of the doubters?"

"Oh, no. He thinks it's bound to work."

Looking up, I wondered how those distant, eternal suns burning in endless darkness might regard ephemeral mortals, who so easily contemplated their own destruction. And did they know how to laugh, those stars, or cry with pity?

Or would they even care?

Equations were scrawled across the dusty blackboard, alongside simplifying diagrams invented by Feynman purely to teach me things only the world's leading researchers comprehended. Einsteinian emergence-matrices measured base-base and gene-gene interdependence; Hamiltonians tracked evolutionary expansion through morphological phase-space. . . .

And a pounding migraine, beginning over my right eye, spread inexorably inward.

"—and then you depolarize the biflange confabulator, and the rekeezy blart destimblefies."

I wrote down half of that on my spiral-bound notebook, then stopped. "Er, what—?"

"I had a feeling"—Feynman was grinning—"I might have lost you there."

I laid down my fountain pen, rubbed my forehead. "There's a faint possibility," I told him, "that you might be correct."

"You're not alone. Einstein hates quantum evolution. With a vengeance."

"At least he understands it."

"I wouldn't go that far." Feynman perched himself on the nearest desk—the room was set up for twenty people, but he had requisitioned it for us two—then threw his chalk up in the air, caught it. "Nobody *understands* the theory. We just know how to use it."

I stared at him.

Feynman's intellect was magical. Whatever he explained became obvious, transformed me into a genius, torrents of energy coursing through my mind. Only later, alone in my room working through cell-function pathways, would I grind up against my own limitations.

"If you told me *why* you need to know this stuff"—gently—"we could focus on what's necessary."

And blow mission security?

But Feynman was correct. I could not manage this without rigorous and *specific* coaching.

I let out a slow breath.

"What I need"—I glanced out the window at sapphire sky—"is to recognize schematics for a nucleic bomb at a glance. Well enough to tell a fake from the real thing, or something that's *almost* a hundred percent correct."

The impish humor stopped dancing in his eyes. "You need to do this quickly?"

"Precisely."

"Under pressure, you mean."

"An awful lot, " I told him.

He mouthed the word *awful*.

"Well, then." Feynman jumped up from the desk, strode into a patch of sunlight, and began vigorously to wipe every equation from the board. "Let's get to work."

"You understand—"

"I've worked out *where* you'll be doing this, my friend." Feynman raised an eyebrow. "You're in the SOE, a real live spook, and you'll be examining enemy plans in situ. Is that how you say it in the King's English?"

"Close enough."

Feynman grinned, but I recognized the shadow that lurked behind his eyes.

"Well, then . . ." In an Eton-by-way-of-the-Bronx accent, "The best of British luck, old chap."

The Quonset hut's interior was dark and sweltering. Exertion threw off heat and humidity as we worked the drills. Palm-heels, elbow-strikes—blinking away salt-stinging sweat—side-kicks to knees.

"Harder!" barked Gunny Rogers.

Panting, my opponent tried for my wrist, but our bodies were slick with sweat—olive-green undershirts sopping wet, dark—and his hand slipped so I thrust up under his chin, swept his legs away.

"Come on, ladies. Let's *work*."

We pushed harder and harder, until we were barely standing, unable to see straight, and Gunny finally called a halt. "Just take five," he said as we breathed hard through open mouths.

In the corner, one of the Rangers was being discreetly sick.

Then I saw Gunny Rogers frown, looking over my shoulder. I turned around.

A big shaven-headed Ranger was throwing a punch—"Catch it!"—expecting his smaller partner to lock the wrist. Except that he came in too fast and hard, the short man failing to grab—

"Damn it," muttered Gunny.

—and the big man swung a long, deceptive uppercut, buried his fist. His victim dropped in a fetal position, gasping.

Gunny stepped forward, but I was closer.

"Actually, old chap," I drawled, "I believe you're misinterpreting." I was still moving as I spoke.

"What do you—?"

Then I whipped both my hands against his big ham-fist and snapped him to the ground. His knees struck hard, pain brightening in his small eyes.

"Captain Fairbairn," I told him, "considers it an *attack*."

And I knew how devastating the wrist-throw could be.

I turned away, trying not to look at Gunny Rogers, who was struggling to hide his broad grin. The man who had been my partner raised his eyebrows and whistled in appreciation.

"You've actually trained with—?"

He stopped, even as I felt the rush of air—*behind*—and I dropped, spinning—

But a big shadow moved past me, Gunny Rogers, and his huge arm shot out and the big man was down. Then Gunny was in the air, boots coming down together in a deadly bronco kick—

"No, Gunny!"

—but he separated his feet at the last moment, boot heels thudding into the mat either side of the downed man's head, raising puffs of dust instead of smashed bones, smeared blood.

"That"—Gunny grinned—"is what you might call the Applegate variation."

When the short, slim Captain Fairbairn demonstrated his combat skills before seated representatives of the U.S. Forces, he tossed a huge bear of a man called Rex Applegate straight into the laps of some very senior officers.

Applegate, at least, had been impressed. He now led the U.S. program teaching deadly combat skills—Japanese and Chinese warrior-techniques, by way of the Shanghai police—to elite forces and covert units.

And, so I had heard, to stay-at-home G-men. The FBI were worried about civilians—demobbed military men, once the war was over—possessing finely honed fighting techniques they did not know themselves; hence their own training. In Britain, we had too much on our rationed plates to worry about peacetime security.

"Not bad, Mr. Brand." Gunny Rogers clapped me on the shoulder afterwards, as we filed out of the Quonset glowing with exertion, strained and exhausted.

"Not bad yourself, Gunny."

Laughter and coarse jokes—some at the expense of the big man who now grinned ruefully: he had learned a lesson—rose up from the others as they headed to the showers. I slipped away, heading toward my small home.

Overhead, the empty early-morning sky gleamed, azure and serene.

Are we primitive animals, spilling each others' blood?

But I wondered, passing along the deserted dusty street, quiet before the working day, whether there were any rational conclusions to be drawn in a world where cosseted intellectuals, civilization's best, with their blackboards and chalk and scribble-filled notepads, could devise modes of devastation far deadlier than teeth ripping artery, blade slicing intestine in the thunder and stink and dirt of battle, and work their own cataclysm of torn DNA and ecodestruction, remotely tearing life asunder while holding themselves aloof from the stink and rawness and fear, at distant remove from the messy, bloody, excremental business of death.

Felix returned two days before my departure, purely to bid me farewell. We stood at the edge of the makeshift town that promised to change so much, and stared out at the dawn-smeared sky, the vast wide New Mexico desert, not needing to speak.

Finally, it was time for Felix to leave. A low-slung Ford, its black-green carapace filmed with desert dust, was waiting for him, a crop-haired driver at the wheel.

"Traveling in luxury," I said.

"Back to Washington, to fly a desk." Felix raised his left arm, watched liquid sunlight slide along the polished hook. "Just what I always wanted, I don't think."

"You'll do all right."

"Sure. I'll be training the neos." With a sudden grin, "They'll be impressed with this against their throats."

"They'll get the point."

"They surely will."

"Well—"

We shook hands.

"You watch your ass over there, Limey."

"Look after yourself, Yank."

He climbed into the back of his vehicle, and nodded as it slipped past me. I watched as it followed the arrow-straight road, black through Martian red, until the Ford grew tiny with distance, was lost from sight.

My final night. Restless, I walked the silvery, moonlit length of Main Street, wheeled left, passed the commander's cabin, Oppenheimer's—

Light.

Reflexively, I crouched. Torchlight flicked across the darkened window from inside, was gone.

A burglar? In Oppy's office?

Raising a hue and cry went against my nature and my training. An alert intruder might slip away. So I moved softly, heel-to-toe, creeping close to the wall until I reached the door.

The knob turned without a sound. Slowly, I passed inside, and closed the door against the night.

Farther inside, the hallway was dark. I crept forward, reached a corner, peered around.

Not Oppy's room.

The burglar had passed on. Light flickered from a cross-corridor. Shadows and darkness were eerily confusing, but surely that was just an ordinary seminar room—

"Oh, very nice." The intruder's voice floated toward me: soft but unmistakable.

What the hell? Feynman?

I shrugged my shoulders once, flexed and released my hands, then crept forward, ready to strike.

But when I stopped at the doorway and peered inside, Dick Feynman was cross-legged upon a desktop, diabolically impish as he played his torch's beam across a blackboard that held nothing that was secret from him.

In fact, I realized suddenly, he was due to attend the lecture that Teller was giving here tomorrow, and it was Teller's own equations that were scribbled across the board in preparation.

I cleared my throat.

"Doing some technical prep of your own, Mr. Feynman?"

He jumped, then slowly smiled, and pointed to an equation. "I can solve this tonight," he said, "in about three hours. I know it took Ed and his pals the best part of a year. But if I do it in ten seconds, during the lecture tomorrow—"

I laughed silently and shook my head. "You don't need to impress anyone."

"I know. It's part of the fun."

After a moment, he slipped down from the desktop, tapped me on the upper arm. "Come on."

Then he led the way into the darkened corridor, as though there were nothing out of the ordinary in being caught sneaking around at night in a classified military installation. Perhaps, for Dick Feynman, it was the merest side note in an eventful life.

I had been right about an intruder in Oppenheimer's office. Feynman bent down at the lock, fiddled with it for a moment, then let the door swing open.

"That was locked, wasn't it, Dick?"

"Not well enough."

"Thank God you're on our side."

* * *

I had been in Oppy's office before, but not in the side room, which Feynman opened. Shelves bore glass display cases—strange shadows shifting in the light of Feynman's torch—and a small lab bench stood at the far end.

"Samples." Feynman pointed. "Don't open anything. We don't have isolation suits."

"Whatever you—"

I stopped, swallowing, and the darkness seemed to sway about me. My body shuddered, as though some outside agency controlled my nerves. I could do nothing but wait out the shaking fit. Soon enough, it subsided.

Hand of glass.

That was what it looked like: glassine, perfectly sculpted, down to translucent bones and sinews and veins within. In other cases, different anatomical parts were likewise near-transparent, some shaded purple as though carved from quartz.

"Preservative stains," said Feynman. "I sometimes take a peek, just to remind myself how horrible our work is. And why we're doing it."

I shook my head, as though to deny the memories made tangible before me. Sweat coated my skin like a new, protective layer; I thought I might throw up. Neither the morphine treatment nor the long sessions with SOE padres had laid those ghosts to rest.

Nor could they ever.

"What's wrong, my friend? You know the kind of work we do."

I was an intelligence agent, trained to remain calm before interrogation, to pass unnoticed through foreign lands beneath the eyes of the enemy. But this . . .

Shards, glinting upon the cobblestones—

This, I would never forget.

"I was there," I whispered, ignoring the tears that tracked down my cheeks. "In Germany, on Kristallnacht."

"Dear God."

I looked Feynman in the eyes.

"God," I told him, "was nowhere to be seen that night."

Memories, made indelible.

It was chilly, that ninth day of November in 1938. I was in Ravensbrück, but the exact location was irrelevant: for on that night, every city, town, and village in Germany became aware of the diabolic movement that controlled the country, of the tiny, scarcely significant number of individuals who were clear-sighted and decent and courageous enough to speak out against it.

The squads wore heavy coats over their brown shirts, but only at first. As the night progressed, they grew warm with their work: dragging Jews from their homes, setting light to their businesses, bringing violence to the streets of Mozart's civilization, to Schopenhauer's culture, and whatever flared up in bedrooms would remain forever spoken only by women's remembered shrieks, by the sobbing of violated children.

Sturmabteilung thugs were everywhere, beating their victims in plain sight of any citizen who cared to draw back his heavy curtains or step outside, while the SA's infection squads screeched up in rented cars and trucks. Those who sprayed the victims were covered in protective gauntlets and hoods, as though unconvinced of their own purity, for the viral complexes were guaranteed harmless to the true Aryan genome.

Helpless, I watched.

Spray-victims scarcely screamed, for their vocal cords were one of the first parts to crystallize, as their panicked lungs sucked virus-laden mist inside and doomed them. When the entertainment lessened, as struggling flesh became static glass and canisters ran out of fluid, the looting and burning started. Eventually, flushed with riotous violence and armed with pickaxe handles, with wood torn from destroyed store fittings, they returned to the corpses, now frozen like glass statues in the street, and put their backs into new work, swinging downward, yelling curses, as they demonstrated their solution to the Jewish Question, the *Judenfrage*.

The next day, as the first groups of prisoners were forced aboard trucks to Dachau, I walked the streets, watching and listening, drinking in every detail of the fearful looks, the shared desperate supportive glee, even as my flesh crawled and every nerve screamed to get away, to forget that Crystal Night had ever happened.

While on the cobblestones, fragments glittered like diamonds in the watery winter sun.

* * *

The transatlantic flight, in the gray condragon's passenger womb, passed in fitful, sickening dreams—featuring Dick and Gunny, above all Laura, in strangely dark, surreal surroundings: good cheer turned to mockery amid chaotic dreamscape—as prep-infusions fought through my bloodstream, reset circadian rhythms, depleted skin melanin to a European pallor.

My escort, the young-looking SOE junior officer who had brought my infusions—were the Special Operations Executive recruiting from the schoolyards now?—said scarcely a word during my periods of lucidity.

It went against the paranoid grain to take debilitating treatment that left me barely conscious, unable to defend myself. But this was not a public flight; I had no right to refuse the treatment.

At one point I woke fully, stared through transparent membrane at the night sky, the silver-capped waves below, and laughed out loud. My junior officer looked worried, but I merely shook my head, unwilling to share the very real memory that had visited me.

Feynman—Dick—had led me into the comms room just before 11 A.M., when we would normally be finishing our intellectual sweat session. I should have realized something was up from the grins the women gave him. As always, glistening black threads filled the room; on them crawled commspiders, pinhead up to fingernail-size, passing between nodelice whose thoraces were swollen with the molecular-encoded fluids which technicians termed *infopus*.

And then, on the stroke of eleven, a sudden phase shift occurred. My neck prickled as half the commspiders rose up on tiny rear legs, raised forelimbs in salute, and I realized the whole cross-and-diagonal formation was the Union Jack, instantiated in web arachnids.

Dick grinned as he, too, saluted.

Smiling, I slipped back into sleep, ignoring the nervous vibrations I was receiving from the youngster escorting me, aware they could mean only one thing: the mission pace was into overdrive; they were ready to send me in.

In London, I passed through Whitehall quicker than the proverbial dose of Andrew's Liver Salts. Final briefing, unusually, was to be in Baker Street; within twenty minutes I was there.

Even though they knew me, I went through full procedure at each of three checkpoints before I was in the heart of the great cubic warren. Formerly Marks & Spencer's corporate HQ, now more concerned with creating codes than selling winter woollies: either way, there was a grim, heartless, lightless efficiency to the place. A gray rectilinear warren, devoid of windows (save for the outer layer, where no secret work was carried out), whose unending sameness made it easy to get lost.

"Fleming . . . How are you doing, old chap?"

"Not bad, Leo. Got a good one for me, have you?"

"Guaranteed unbreakable."

"Or my money back?"

We both laughed.

But that was only codes-and-ciphers; my briefing officer for the mission per se was a nervous-looking man with slicked-back hair and heavy glasses. His dark suit looked two sizes too big for his scrawny shoulders.

I did not know him well, but his name was Turner and his reputation among the field agents was of a ferocious intellect, second to none, whose planning paid as much heed to getting his people back as the initial access phase. They loved him for it.

But I knew that the powers-that-be used Turner sparingly, for the high-risk projects where potential gain outweighed perceived danger, in the opinion of Whitehall's desk-bound analysts.

He poked his head into the anteroom where I was waiting. "I'm ready for you now, old thing."

"Civil of you." I followed him into the windowless room and sat opposite his desk. "Lewis would've kept me kicking my heels for fifteen minutes, just to remind me who's boss."

"Ah." Turner removed the spectacles, which I wasn't sure he needed, polished them on his wide silk tie, replaced them. "That's an Old Harrovian for you."

"You'd be an Eton man."

"Not quite." With a sly smile, "Barnsley Grammar School for Boys. On scholarship."

I wondered what Feynman would make of the social minefield that was our all-nurturing but ruthless class system.

"You could have fooled me," I told Turner, and hoped he would take it as a compliment.

"Language check." He steepled his fingers, getting to business. "You have fluent Polish?"

My personnel file lay on the desk, unopened.

"Rusty." This worried me.

"You don't have to pass for a native," Turner assured me. "You *will* have a German-national cover-ID, but I'm aware that's no problem. You can assume regional accents?"

I nodded, knowing that the real briefing was about to start.

"Very well. This—" Turner laid out a series of wide glossy black-and-white photographs: dark forest, slate-colored shore, choppy ocean waves. "—is your target area. That's the Baltic Sea. And these men form your objective."

Two individuals, three photographs. Clear portraits of men in formal suits and bow ties; both of them scientists, I knew. Classic Aryan profiles.

"Code names Wilhelm One," said Turner, "and Wilhelm Two. Their identities are hidden within the Reich, but they do in fact share first names. And the photographs, as you can see, postdate morpho-surgery."

"Final cut?" I needed to know if their appearances might have changed since then, to move even closer to the Reich's cosmetic ideal.

"We think so. They're too busy for further surgery."

In the third photograph, taken upon an airfield with a dragon squadron as backdrop, "Wilhelm One" was standing to the rear of a group dominated by Hermann Göring. Though the scientist was mostly obscured from view, his black SS uniform was unmistakable.

"Why two men?" I asked, knowing it doubled the risk.

Whistles. Men's shouts echoing in alleyways. Killer dogs, barking.

I shook away the memory.

"This one"—Turner pointed at "Wilhelm Two"—"is designing what flyboys call the payload. His institute in Berlin is carrying out the same work as our friends in New Mexico."

"You're sending me into *Berlin?*"

His gaze, dishwater gray, fastened on me for a long moment, and I wondered whether my show of nerves had blown the mission before the start.

"Access and egress," he said evenly, "would take too long. Not to mention the other difficulties. Wilhelm One, on the other hand, is developing long-range nymphcluster dragons on a genetically isolated development site in Usedom."

"All right. The Baltic, then."

Breathe. Stay calm.

"And this chap, Wilhelm Two—Willi Zwei, we've begun calling him—is the one who appears to have made contact with us. I believe it's an insurance policy, in case his side loses the war. At the moment, it's a policy he probably thinks will never pay out."

Most Nazis would stake their life savings on all-out victory right now. These background ambiguities scared me.

"We're not sure"—I wanted to be clear on this—"who made the contact? They came to us first?"

"Not exactly. Polish Intelligence had already alerted us to the existence of this development base." Turner tapped the forest photograph. "The SS might have become aware of the partisans' surveillance, and set up a scam."

"Dear God."

"If I thought this was a likely scenario"—with frost in his tone—"I would not be sending you in. Is that clear?"

"Understood."

"Willi Zwei is due to visit the base from Berlin. I expect you and your team on the ground to make rendezvous with whichever Wilhelm is behind the message—"

"They made contact via a cutout?"

"Of course. But the content indicates it must be either Wilhelm One or Two: they passed on info only one of them could have known."

No need to ask for details. If they were pertinent, he would tell me.

Turner talked me through the scenarios; none of them was pleasant. The worst possibility was that their project was stalled or fake, and I was falling into a trap specifically designed for someone with enough knowledge to verify nucleic bomb schematics at a glance.

"That's all," he said finally. "Thomas Cook's will see you through the rest."

Their official designation was Clothing, Travel, and Firearms, but we called them Cook's Travel Agency because they sent us to such exotic places, where we could meet interesting people and with any luck not have to kill them.

Quite often, we came back.

"Thanks very much. I'll send you a postcard."

I tried the operator again.

"Whitehall one-four-nine-eight," I told her.

Once more, the empty ringing tone.

Before I left the Baker Street cube, Leo called out to me. He was waiting at the intersection of windowless corridors that I was most likely to take, grinning while a uniformed sergeant gave directions to a newly joined-up FANY who was obviously lost. She frowned prettily, unused to deciphering Geordie accents.

"Well, pet, first you go left—"

"Come on," said Leo. "There's someone you ought to meet."

I followed him back to his office, but it was empty. Leo surprised me by jamming a slightly disreputable trilby upon his head, then shrugging on his overcoat.

"We're going to Victoria."

"How's she doing, anyway?"

"Grow up, Fleming, why don't you?"

As we walked along Baker Street, it occurred to me that sandbag manufacturers, to judge from the dirty piles around every doorway and window, were making a fortune. I looked up; the skies were clear of all but cream-and-gray cloud masses. It would be nice to see some rain.

"How're your FANYs doing, Leo?"

"The resend rate is cut right down."

And agents blessed him for it. X-radiation checkpoints, sporemists, and natural hazards garbled many cryptocytes, whether avian-borne or carried by couriers beneath fingernails, or secreted in places it was best not to think about. If Leo's teams could decipher a part-randomized base sequence from the field—in effect, cracking a code that was not the one the agent had intended to use—it saved a second dangerous journey, with a vastly increased risk of capture.

"So who are we going to see?"

"Someone you'll like, I promise."

By this time the evening was growing cool, and we walked faster as we turned into Victoria High Street. Then we stopped opposite Westminster Cathedral, with its oddly pleasing mixture of brick-red and gray, and the bicolored tower rising upward. Leo headed across the road, and I tagged along.

Inside, the cathedral looked soot-blackened but darkly impressive. I followed Leo through a discreet door at the side, nodded to a priest in vestments, and climbed a creaking wooden stairway. At the top, Leo rapped on a door.

"It's me."

"Come in, you."

Inside was a small oak-paneled office whose shelves were crammed with books. A tall redheaded man rose from his bureau to greet us.

"This is Jack," Leo told me. "And I'll warn you now, he's a Jesuit. Don't expect to win any arguments."

I held out my hand, smiling. "I won't. How do you do, Father?"

"Jack. Please."

Leo peered at the shelves. "He's also the best geneticist we have."

"Second best." Jack waved me to a wooden chair. "You're too modest, Leo."

"I don't think so. There's something I could use your help with—"

As they launched into an arcane discussion of phage-borne randomizers, I walked over to the shelves that Leo had been examining. A small plain crucifix hung overhead, facing the picture of the Sacred Heart. Beneath lay a crammed mass of titles: Wallace's *Descent of Man;* Schrödinger's *Life Waves;* the complete Gibbon's *The History of the Decline and Fall of the Roman Empire*. Everything Immanuel Kant had written, in the original German. *Gray's Anatomy*. Hobbes's *Leviathan*. A slim volume of *Emergenic Transforms,* by Riemann. Machiavelli's *The Prince,* in both English and Italian. *Tarzan of the Apes,* by Edgar Rice Burroughs.

Open on the desk, beside a cheap brown-robed statuette of some saint, was a copy of von Clausewitz's *On War*.

"Interesting," I murmured.

"Saint Gregor Mendel?" asked Jack, mistaking my interest. "He's one of my heroes."

"Of course." I wondered in how high a regard he held von Clausewitz.

"Have you ever wondered how different the world might've been, if the Vatican legate hadn't plucked him from obscurity and spread his work across the globe?"

"Um . . ."

"I'm afraid the Philistines have arrived," said Leo.

"Story of my life." Jack held out a small glob of blue gel toward me. "Here you are. It's a new antiviral. Best protection you can have."

So he knew I was going into occupied territory.

"Thanks, Padre. I mean Jack."

At that point, Jack broke out a bottle of Beaujolais, which he assured us had not been consecrated, and we drank a small toast or two and chatted about everything from girls—Jack laughed a lot—to Kant, to the war in Africa.

It was only later, at the end of our convivial evening when an SOE driver came to collect me and Leo, that I wondered what was really going on. Had Leo intended to cheer me up, the day before I went into danger? Or had he set up this distracting soiree purely to keep me busy?

Chatting beside me on the rear seat, Leo looked too fresh and innocent for the deceptive world in which we moved. But, even as I responded to his questions, all I could really think of was the one person who had come to matter, my small core of stability while everything around us slid into chaotic uncertainty.

Night flight.

Crouched in my transparent half-shell, ready to roll—

"Thirty seconds." The jumpmaster checked me, nodded.

A gunner peered back inside the long fuselage, grim-faced. Probably wondering whether I was worth it: there had been ack-ack on the way in, and Pterafighter or Falcon squadrons were probably already rising; the flight back was not going to be fun.

As for me . . .

"Ready—"

Slipstream, as the hatch furled back. The jumpmaster reached toward me.

"—Go!"

Then darkness and buffeting winds were all around, fear shrieking inside, as the half-shell completed itself, became a cocoon.

Darkest night above me, and below. Tumbling . . .

And drop.

Wet ripples on the silver moonlit grass: the drop sphere and its chute membrane were half-dissolved at the foot of a night-black hedge. I waited . . . then a soft whisper sounded, and I dropped. Shadows against the night: four men, maybe six.

Wide-eyed, breathing fast and shallow, I crouched like a sprinter at the blocks, scared as hell and ready to explode into—

"Hello, darling." Liquid words, sliding through the chill night air, struck straight through my trained defenses, and I nearly choked with shock.

Laura?

I should have realized. Even the initial infiltration, the setup, would have needed specialized knowledge.

"Dzien dobry . . ." A male voice. I could just see the rifle outline in his hands.

"Miło mi pan poznać," I told him.

But it was the figure at the group's center I wanted to see.

Laura's here!

A week of dream and nightmare: of days holed up in safehouses—here an attic, while German voices rose up from the street below; there a deserted farm, the farmer's ruddy-faced wife giving cause for concern as she slipped away unannounced: we broke cover and made our way into the woods, just in case—and of the nights, traveling through heathland, which became steep slopes blanketed with thick forest, filled with the rustle of tiny creatures, wind brushing branches, and the surprising hoot of a predatory owl.

And Laura, my Laura, was with me.

"You're good at keeping secrets, my darling," I told her. "I should've guessed why Leo kept me occupied in London."

"There's no secret how I feel about you, dear man."

Kindred spirits, and more.

When we traveled at night, we walked as one: each holding the other's hand, sometimes with our clasped hands in her coat pocket or mine, for warmth. Among the four Polish agents with us—who moved through darkness like pale, alert cats, their night vision enhanced even beyond mine—there was a calm acceptance of the relationship, with neither joking nor resentment.

They knew how easily lovers could be wrenched apart by the capricious, devastating daily tragedies of war.

Finally, after a cold few hours spent crouched beneath undergrowth while roving scanbats passed overhead—dark flutterings against a strangely golden moon—we trekked on through winding forest paths, finally descending to the town of Lubmin.

From the forest's edge, we watched. As false dawn smeared the eastern sky, stars glittering against a backdrop thickening into navy blue, our small group stood watchful, breathing pine-scented air, regarding the cobbled streets and peaked roofs. There were no patrols.

And then a window shutter moved, just beneath a wooden eave, and someone laid a striped dishcloth across the granite sill. That was our signal.

One by one, we slipped across the open ground with nerves screaming against a hail of bullets that never came, boots in hand so that we walked on cold uneven cobblestones in silent socks, reached the back door leading to the safehouse's scullery, and slid inside.

The first morning was the worst.

In a threadbare coat and dirty shawl, head bent forward, Laura shuffled into a group of workers on the gray cobbles below, while I watched from the attic window. Then a dark green flatbed Volksnashorn transport pulled up, and one of the Wehrmacht soldiers jumped down, jackboots clattering on stone. He rapped the armored hide in a control sequence; the rear unfurled, formed a ramp up which the workers could drag themselves, ready for another long shift at The Keep.

I watched as they drove into a narrow street where houses leaned inward, took a sharp turn at the far end, were lost from sight.

She's been doing this, I told myself, *for months.*

But even a long-established cover could be broken at any time.

For the rest of the day, nervous in my attic and unable to go out, I drove myself physically: stepping up and down from an old sturdy apple crate, performing sit-ups and press-ups, chin-ups from the rafters, striking empty air as I worked through the killing techniques of Fairbairn's system.

And then, when the nerves were finally quiescent, I lay prone on the floor with an imaginary long gun held in my hands, and mentally rehearsed the shot I was going to take. This visualization, entirely serious, was the only practice I would get.

When evening came and Laura climbed up to the attic, I took her in my arms, hugged her as though I could never let go, feeling the coldness of her skin, inhaling the faint laboratory scents clinging to her hair and clothes, too overwhelmed by her safe presence to speak.

She kissed me then, an explosion of warmth, and I abandoned thought to the immediacy of the moment, when communion needed no words.

It was on the fourth day, as I waited in the overfamiliar attic—dusty pale-amber collimated sunlight tracking the morning's progress across gray knotty floorboards—that a door banged below, and I jumped.

I waited for Elsa, the plump red-haired woman who owned this place, to call a greeting. Then softer footsteps rose upward from the ladderlike stairs, and I crouched ready for combat—

"It's me."

—then relaxed, recognizing Piotr's voice. He was the team leader, whom Laura called Petya, but I used his proper name. The others—Zenon, Stanisław, and Karol—I had scarcely spoken to at all.

"Hello, Piotr. Everything okay?" My voice came out more stressed than I had intended.

"So far." His lean pale face betrayed little emotion: a face that had seen too much. "Everything is in place."

"Ready to go now?" My skin shrank.

"In a few hours. You're sure about the long shot?"

I understood: Piotr and the others were hunters; they did not trust me to make the kill.

"I placed second," I told him, "in Bisley, before the war."

"Bis—?"

"National rifle championships."

"Ah. Is good."

For Laura's sake, it had better be.

"You rest now," added Piotr, "and I call you when ready."

"I'll do my best."

Later, as the summer evening's sky finally darkened to gray, there was a knock from below, and Piotr's voice again: "It's time."

People moved warily in the quaint streets, where inward-leaning houses with uneven leaded windows appeared to stare down at dark gray cobbles. Hunched shoulders betrayed the locals' awareness of constant observation: from spybats, from passing Waffen SS patrols, from each other. The landlady who wanted a new tenant; the disaffected pupil hating his teacher; the jealous neighbor bottling up resentment over the years: anyone might turn informer, spill out their suspicions to the *Sicherheitsdienst,* convince themselves of their countrymen's disloyalty.

And would they stand and watch afterwards, as their victims were taken away by stone-faced squads whose self-righteous force could not be denied? Or might they never stop to think of the suffering about to be inflicted on those who fell victim to the Reich's internal guardians of pure Aryan thought?

We're all products of the culture we—

I halted, just for a second, at the sight of gray uniforms ahead.

Don't stop.

Piotr, walking ahead of me, continued without reaction: showing more professionalism than I.

"Papiere."

I ground to a halt, my guts tightening.

"Ihre Papiere." Hand outstretched, impatience entering his tone. *"Schnell."*

But it was his comrade's hand I noticed: flexing, itching to whip up the Mauser and squeeze the trigger for that momentary burst of flesh-destroying pleasure.

"Bitte." I handed over my ID.

We used to say bad things, we agents, about Thomas Cook's, and the fussy matrons and old men who ran the section, but the documentation they produced was usually—

"Also gut."

He handed them back, and *this* was the greatest danger: that I would reveal overwhelming relief, as my protective facade fell inward, no longer shored up by fear tension; relaxation would be my downfall.

For a moment I thought the other soldier was going to fire, and I prepared to claw at his eyes, to take some recompense with me to the grave—

But they were walking on now, not quite in step, and the sickening realization of my own mortality was washing through me in waves. For now, I was safe.

A scuffed footstep—Piotr, turning a corner up ahead—brought me back into reality. Here security was ephemeral, a guarantee of survival only until the next flashpoint, and I had better get a grip or we would be done for.

That included Laura.

Shards, on cobblestones—

For Laura's sake, I hunched my head forward, tucked my false papers back in my too-thin jacket's pocket, walked on.

From a thicket beyond the town's edge, we stopped, stared back at armed curfew patrols heading into the alleyways, striding with a centered arrogance among ordinary people, knowing they could burst into homes upon a whim, always able to justify the violation of frail civilians in the name of the dark power that ruled their once-civilized selves.

A soft rain, like silent weeping, began to fall as we turned into the darkening forest, moving quickly now.

* * *

And then it was time for the long shot.

Damp grass lay beneath me, but the cold faded as I pulled the hard butt into my shoulder, sighting the base of the tower that guarded the narrow defile, and began the breathing ritual. In the crosshairs, the sentinel was visible: a paler shadow inside his darkened booth.

"The other sentries," whispered Piotr, "are almost out of sight."

And breathe . . .

Behind me, though I could not turn around, I could sense the warmth of the other three Polish fighters, crouched in readiness.

"Three. Two. One. Gone."

I had twenty seconds exactly.

And hold.

Sweat on my finger, around my eye.

Ignore.

Centering the target's image—

Hold . . .

A wavering . . . Then steadiness.

Now.

Squeeze.

A shift in image, shadows in my sight, unable to tell if—

"He's down."

Success.

It was an airclaw, silent and accurate: a sniper's bullet. No ganglia to be disrupted by defensive sonic fields; no gunpowder whose crack would pull a thousand Wehrmacht troops and Waffen SS down upon our heads.

The rifle's weight lifted as strong hands tugged my shoulders.

"Now go."

I was on my feet and moving.

The five of us ran like wraiths, up the slope with lungs and thighs burning in the cool night air, fast and silent. Beside me ran Stanisław, the tallest of the Poles, carrying the heavy rifle one-handed, keeping pace despite the Browning's weight.

When we reached the tower, Stanisław dropped back and headed into the sentry's booth. He would strip the body and don the uniform.

"Two minutes," Piotr reminded him.

A change of guard had just occurred. From the info Laura had provided, the next perimeter patrol would not expect to recognize Stanisław as the sentry on duty: from a different unit, newly posted to this place.

"No problem. Just go."

We moved on, into the narrow defile, cloaked by darkness at the rock face's base.

Rumble. Stink of sulfur.

Disgusting.

Hot breath played across my skin as the bulky hyperkomodo sniffed, licked my sweat with its reptilian tongue, then turned and lumbered away into shadows. Beside me, Piotr sagged: he had not been convinced the pherocipher would work; but we smelled like friend, not foe, to the ultrasensitive beast.

Then a wide flatosaur transport came into sight, moving along the defile with its scaly hide almost scraping the rock on either side. Counting carefully, I took a breath and then rolled between two massive legs—it had twenty four in all—and whipped my hands up, hooking fingers desperately into ridged plates as it dragged me along the broken ground, and I swung up one foot—*missed*—then got the hold, pulled up the other, and then I was clinging on by all four limbs, like some desperate parasite hoping not to be shaken off.

When I could spare a glance, there were three other primate shadows—Piotr, Karol, and Zenon—splayed against the underside with me, holding on while the lurching transport carried us out into the heart of the enemy's installation.

Through the internal checkpoint. Half hearing the driver's chat with the guards.

"—geht's mit dir?"

"Ausgezeichneit. Arnold hat ein neues Mädchen—"

Then we were past, and I let go my hold.

Drop.

Stone thudded against my back. I timed the massive legs' movement, counting—

Now roll.

In the small courtyard, a wooden doorway swung open, and a blue firefly glimmered for an instant before a small fist hid it once again.

Laura.

I was first into the narrow hallway, and I brushed Laura's cheek with my fingertips, but there was no time for any other greeting as we moved inside. The walls were lined with polished mahogany, from what I could glimpse in the firefly's blue-tinted illumination.

Then, at the hallway's end, we turned into a cross-corridor. This had been some landowner's grand house, but as the corridor's walls became bare stone and the air temperature dropped, it became obvious—even before the floor sloped downward—that we were heading *into* the low craggy ground, which splayed out, forming a headland, into the cold Baltic waves.

There was a weapons area filled with soldiers, swarming across the equipment, while officers barked out commands and a black-uniformed Gestapo colonel oversaw the activity. Ducking low, we used a wide pipeline for cover, passing into the barracks area—more soldiers, dining—then through an internal pherolocked door which responded to Laura's touch, into the cavernous pits.

A small catwalk, of wet-looking black chitin, extended bridge wise across the pits, to an armored megarhino-hide door on the far wall. On either side of that door, in small recesses in the raw rock face, a big handle shone a dull, fluorescent red.

Laura leaned close to me. "I'll go first, my love."

Her lips brushed my cheek, and then she was moving across the catwalk, in plain view of any sentries who might make an appearance below.

I held back, letting Karol and Zenon follow her—it would be their job to hold the deadman switches, keeping the armored door open—before making my own way onto the black chitinous catwalk, with Piotr close behind.

Small phosphorescent cocoons, adhered here and there to ceiling and walls, provided a hellish light. Far better, despite the moans that rose toward us, to keep the pits' poor dwellers hidden from human sight.

It was worse than Kristallnacht.

For the things that moved beneath me now, the once-human beings that fluttered and crawled and flowed in the cold pits, dragged slime-trails across broken rock or simply melded with it, were inflicted with something worse than agonizing death: a tortured ongoing pseudolife of glistening flesh and ever-raw wounds, of strange sprouting limbs—many-jointed protrusions, boneless ulcerated tentacles—and weeping, putrescent growths in an ongoing shamble of sickening flesh that could not die.

Behind me, a sharp inhalation, and a whispered name: *"Brigitte . . ."*

Piotr, ashen-faced, was staring down at that jumble of animated meat, where a few strips of cloth—there, just visible, a dirty yellow star—remained, caught in folds as the hypertumorous growths had split clothing asunder. On a liquid protrusion that might once have been a head, a stretched mask, like a face impossibly distorted, pulled down and to one side, opened its pseudomouth in a silent plea, or recognition.

I caught Piotr's arm. "The only thing we can do," I told him, "is avenge her."

A tense nod, sinews standing out like cables on his pale neck, and then he followed me.

But he looked back as he walked, keeping that tortured facsimile of life in view for as long as he could, until we reached the armored door.

Karol went to the right, Zenon to the left.

"Jeden, dwa, trzy—"

Simultaneously, each leaned down on a deadman handle. The door puckered, then slowly furled back, revealed the steel-lined passageway beyond.

Karol and Zenon would remain in place, for if either man released his hold, the doorway would close. It could be opened normally only from the outside. Unauthorized egress was possible, in case of emergency; but using the interior handles would set off every alarm and klaxon in the installation, shutting down our mission in a matter of seconds.

"All right," said Laura. "We've made the rendezvous."

Let's hope Wilhelm manages the same.

We moved into the steel passageway.

Lab benches covered a factory-wide floor. Laura, Piotr, and I walked the length of one aisle, to the flight of pinewood steps at the far end, where we stopped. The steps led up to a glass-walled cabin on a raised level, but the cabin—an isolation lab, perhaps—was unlit, as deserted as the ghostly expanse of benches, unoccupied stools, and workplaces ranged across the wide, gloomy main chamber.

On one wall, high, hung a scarlet swastika-emblazoned banner. Below it, the only other door into this place slowly swung open.

"No . . ."

And *two* blond-haired figures, white lab coats over their impeccable suits, walked inside and stopped.

For a long moment, we stared at each other, locked in silence.

The contact, via a cutout—a local schoolteacher with little knowledge of the Nazis who had contacted her, or of the resistance fighters with whom she engaged purely in writing, via a long-established letter drop—had evinced knowledge of the long-range dragon program, and the nucleic bomb payload that their nymphclusters were designed to deliver. Whitehall's analysts had been certain that one or other Wilhelm was involved: no one had considered the possibility that *both* of them might be considering defection to the enemy. Especially now, as the Blitzkrieg was pulverizing England's cities, now that jackboots marched throughout Europe, once-free countries having fallen like dominoes before the encroaching Reich.

I don't like this.

It was too rich an offering to accept at face value.

"Let me show you something." The shorter of the two Wilhelms—Wilhelm One—gestured toward an opaque pearly panel upon the nearest wall. Speaking in English again, "Come here."

There was no point in shyness. I walked over to him, intending to offer my hand in greeting, but he turned away and dialed the panel into transparency.

Automatically, I moved to one side, but the lab surrounding me was shadow-shrouded, while the great hangar chamber into which we looked was brightly lit with white incandescent arc lamps. If any of the ground crew looked up, they were unlikely to spot us.

"Dear God," muttered Piotr.

For we were looking down upon the hugest dragons imaginable, their vast delta-wings filling the great shallow pits, stretching from wingtip to wingtip as wide as a row of houses. Nymphs, opalescent and spherical, their long tendrils neatly bundled up around great yolk sacs, were being loaded into the dragon's cavities; once released over the target city, the ejected nymphs would blossom, shed their skins to become small adult dragons within seconds, swooping down in operant-conditioned precision to deliver their deadly payloads directly onto the soft civilian targets below.

"The first test flights," said Wilhelm One in an even tone, "have proved satisfactory."

"What's their range?" I asked. And, when he did not answer, *"Was ist die Fliegweite?"*

"Zwei tausand Kilometer." He shrugged. *"Vielleicht mehr."*

With a two-thousand-kilometer range, they could launch from here—would not even have to relocate their facilities to occupied France, as I had expected.

Even without the nucleic bomb, if the Luftwaffe labs could spawn these hyperdragons fast enough, hatch them in sufficiently large batches to provide a dozen squadrons, maybe twenty, then the war against England would shortly be over, and Englishmen would be learning German in the same way our Saxon ancestors were once coerced into adopting Norman French. But the new regime would be something out of nightmare, such as the writers of medieval epics, for all their monsters and demons, could never have imagined.

"And you." I turned to face the other Wilhelm. "Just what are you hoping to get out of this?"

* * *

They wheeled blackboards on casters into place, near the pinewood steps that led to the isolation lab. Quickly, Wilhelm Two scrawled equations in chalk while Wilhelm One, readying a glass dish upon an asbestos mat, gave a commentary.

"From what I understand—" He waited until I nodded: I knew he was not a specialist in quantum molecular evolution. "—there are two problems. First is the bomb core itself, with self-replicating attractor-strands designed to cascade through the atmosphere."

I felt myself grow cold.

"Second," Wilhelm One continued, "is the vector trigger, which energizes the process and carries the spawn outward, enabling the cascade."

Even from party talk in Oppenheimer's house, I could have verified this much: that the Nazis were a long way down the line, to have gained insights into the trigger mechanism, to have realized that it was as crucial to the weapon's operation as the nucleic core itself.

From the enzyme formulae currently growing upon the board—Wilhelm Two tossed one worn nub of white chalk aside, picked up a fresh stick—I could see, also, that they had at least the basics of core-construction techniques. Just as (contrary to Einstein's instincts) characteristics like intelligence had evolved in the natural world too fast for strict Wallacian macroselection, so, too, could Wilhelm Two and his coworkers produce the necessary replicator strands in far less time than the millennia that ordinary evolution would require.

As Schrödinger and Bohr had shown, enzymes act as chemical observers, forcing the collapse of molecular wave-functions—overlaid simultaneous possibilities—into one outcome. But an enzyme that repeatedly "observes" a molecule in the same position prevents it from once more entering a fuzzy, overlaid state, essentially freezing its configuration and energy forever.

But quantum biology had another, counterintuitive tenet: that the way in which a measurement is carried out partly determines the value which results.

And the New Mexico scientists made use of this. If the observing enzyme itself was slowly changing, then *it forced the observed molecule to evolve* in ways determined by the enzyme's own biochemical propensities.

"So what," asked Wilhelm Two, turning from the blackboard, "do you think, my friend?"

I tried to keep my face a frozen mask, but the Wilhelms exchanged a glance, and I knew that I had betrayed too much.

Whatever the results they had obtained so far, I knew now that the Nazi nucleic bomb project was a viable program, and that if they continued long enough, they would succeed in destroying any ecosystem they targeted, including human life.

"Also gut." Wilhelm Two nodded. "Good enough. Let me show you—"

A distant door clanged, and both Wilhelm's faces grew pale simultaneously.

They were not acting.

"Patrol," said Wilhelm One.

Obviously unscheduled.

"Quick." Wilhelm Two pointed up at the isolation lab. "Get inside."

Piotr reached inside his jacket.

No—

He must have hidden a revolver there, despite my orders. The idea was, no firearms to be brought inside the installation itself. If we had to shoot, we were dead, our mission a failure.

"Okay." Piotr looked at me, then took his hand out of his jacket, empty. "I agree."

Laura went first up the pinewood steps.

"Hurry."

Inside the isolation lab, Wilhelm Two pointed at two boothlike reaction cupboards built into the rear wall. He quickly pointed from Piotr to me.

"You two will, um, *Ihn verbergen . . .*"

"Hide," said Wilhelm One, staring through the window to the wider lab below.

Laura stared at me.

"No," I said.

But then, "You must." She reached up, brushed her fingertips across my lips. "There's room for you two, and I'm on the staff roster—"

"Quickly." Piotr grabbed my sleeve. "We must."

"All right."

Twisting a Bakelite control-knob, Wilhelm nodded toward the transparent-fronted cupboards. The outer membrane slowly darkened to black opacity.

"Squid-ink derivative."

"I—"

But the doors outside were opening for the patrol, and there was no time to chat as I pushed myself inside, onto the waist-high surface, and pulled myself into a hunched position as the membrane hardened, trapping me inside the reaction cupboard.

Laura—

But I could see out into the isolation lab, with Laura shivering as the patrol burst into the shadowed area outside, as the two Wilhelms briefly conferred. Sounds were muted, the words indecipherable, but the ink-soaked membrane was transparent from my side.

I let out a sound that was half a sob, then held myself still, breathing fast.

The isolation door opened, and an SS officer stepped inside, his black uniform magnificent with silver badges and decorations. His round face looked almost pleasant, save for the thin purple scar and the lifeless gaze that passed over the cupboard membrane—I shivered, unable to stifle the reaction—and continued on to Laura and the two Wilhelms.

"—Sie hier?" His voice raised high enough to penetrate my hiding place.

Wilhelm Two muttered something, of which I could decipher only the word *arbeiten*. Claiming that they were working late on something important.

Did the SS distrust their own chief scientists this much?

Behind the officer, two gray-uniformed troopers took up position inside the doorway, hands upon their Mausers. But it was the SS man, with the fat white baton inserted in his shining belt, who scared me witless.

Laura, my love.

For a microsecond, her eyes flickered in my direction. Then she turned away and looked down at the floor as she answered a direct question from the officer.

NO . . .

I saw then that she was holding the glass dish from the lab bench where Wilhelm One had been standing; the scar bunched on his face as the SS man, too, noticed it. Off to one side, a small red light was blinking, and I wondered if Laura's taking the sample dish had triggered some kind of silent alarm.

The officer barked a question, but did not wait for Wilhelm One—already, to his credit, stepping forward—to answer.

Instead, the baton slid from the black leather belt, its tapered nozzle just inches from Laura's fine face—

No! Laura!

And I clawed the membrane and might have yelled, but it was already too late as the sporemist squirted and the woman I loved was a zombie before my fingers struck the membrane.

I stopped, frozen, unable to comprehend what I had just seen.

One of the troopers opened his mouth, about to speak—he might have heard me—but then he caught sight of the SS officer's creamy smile and subsided. He exchanged a glance with his comrade: I could see their unspoken decision to remain quietly inconspicuous.

Laura. Oh, my Laura.

Too late, now.

Expostulations, explanations. I could not care whether the Wilhelms cracked and gave me and Piotr away, or if they all walked away and left me here to suffocate and die. For Laura, the standing corpse of the woman I loved, was there before me: separated from my desperate grasp by a quarter-inch of impenetrable membrane and a lifetime of devastating regret.

When the membrane finally dissolved and hands helped me out, I could scarcely process the information that Wilhelms One and Two were here, that Piotr's hands were fastened on me like iron claws, and that the SS officer and his patrol were gone.

"Laura . . . ," I said, to the one person who could not hear.

Who would never think a human thought again.

I reached out to touch her cheek, but held back from touching, in case of infection: a moment's cowardice that will never leave me. But there were tiny pustules already sprouting across that once-flawless ivory smoothness, precursors to the gross transformations that would soon render her fit company for the tormented monstrosities incarcerated in the pits outside.

And was that a glimmer of tortured awareness in those fine, dead eyes?

I hoped not. I prayed that she was truly gone.

Oh, my Laura.

Wilhelm One cleared his throat then, and said, "I am very sorry, sir."

Looking up, I saw him swallow, the same awareness registering in Wilhelm Two's face: the killing rage, the cacaphonic roar inside, inciting blood-vengeance in a wave of eye-gouges and throat-strikes. I could destroy them, tear them limb from—

"My friend." Piotr drew his revolver. "If you tell me, I will shoot them now."

The gunshot would be suicide, but I could not care. And the two Wilhelms, swallowing, dared not speak.

But finally I tore my gaze away from the one who mattered, and said, "What do you want? Really want?"

"You approve the bomb plans?" asked Wilhelm Two. "They are acceptable?"

"I—"

And *that* was the moment when the mission—and the world—could have come crashing down. Because there was another blackboard in here, another chalk diagram annotated with Wilhelm Two's scrawl, and its similarity to something Dick Feynman had shown me brought me to a halt.

"I'm working on something," he had said, *"called Quantum Evolutionary Determinism. It's years from completion, but there's a technique you might use—"*

And so I used it now, in my head: Dick's integration-over-all-futures technique, as I stared at details of the enzyme whose job was to force-evolve the replicating core, and I knew now it would never work.

Before, I had given away my thoughts. But with emotions burned away by Laura's zombie-death, I had no reactions to betray me. It saved me, now that life was no longer worth anything at all.

"—think they're fine. Whitehall would love to have you."

"But I would need to continue my work."

"You'll have a house in Oxford," I told him. "A big one. And you'll work with the finest minds we have."

A lie. Our scientists were not based in Oxford: too obvious a target.

"Very well, then," said Wilhelm Two.

A decision crystallized in Wilhelm One's eyes, and he nodded abruptly to his namesake.

"Good luck, my friend."

Then he wheeled on one heel and walked quickly from the isolation lab. His footsteps clattered back from the pine steps, clacked across the polished parquet flooring, and then he was through the side door beneath the scarlet banner, and lost from sight.

"I guess, my friend," I said to Wilhelm Two, "you'll be coming with us."

But there was one last development. Piotr, with a bottle of reagent in his hand, which might have been hydrogen peroxide, pulled a labcoat from its hook behind the door. He put down the stoppered bottle and hauled on the white coat.

"You two get out of here."

"I—"

"The patrol might come back. If they don't look closely, they'll think"—with a gesture toward Wilhelm Two—"that I'm him."

I answered Piotr, though I could look only at Laura as I spoke. "All right," I said. "We're going."

And, leaning closer, *"Goodbye, my sweet darling."*

Then I walked away without looking back, leaving Wilhelm to follow me if he chose.

When we reached the far end of the steel passageway, Karol and Zenon were there, still leaning on the deadman-handle door controls, looking at us curiously.

"Where—?"

Then a shot banged out, loud and flat, from the labs behind us.

Laura!

And I knew that Piotr had granted her rest in the only way possible.

"Come on." I took hold of Wilhelm's sleeve. "We're moving fast now."

For his own sake, he obeyed the implicit command, not knowing that I saw through the sham, would not reveal my knowledge of the program's uselessness.

Because if I did, I jeopardized everything: the fate of Britain, of the western world. Betrayed the cause that Laura had died for.

Oh, my love.

Saved the world for which I no longer gave a damn.

There were rushing patrols inside the grand house proper, heavy boots thudding on the carpeted hallways as they rushed to grab their weapons. But they allowed the senior scientist through, away from any danger that might have broken out inside the labs.

We were too far away to hear more gunfire, but I knew that Piotr would not sell his life cheaply.

Then out along the narrow defile, whose far end was marked by four bloody corpses—an entire patrol, dead—and our comrade Stanisław, mortally wounded, white-faced and whimpering, inside his sentry box.

Karol stepped inside with him, murmured comforting words, then used a Wehrmacht dagger to grant the only absolution possible.

I was too numb to feel anything as we trekked across open ground to the forest, to begin our long journey to the rendezvous point and freedom.

There must have been a flight, an RAF pickup; but posttraumatic amnesia set in, they told me afterwards, for the reminder of the journey home remains forever lost in vacuum.

Little of the debriefing, in an isolated Wiltshire farmhouse, comes back to me, either. But I recall the conversation I finally had with the Old Man, when I returned to duty in Whitehall.

"Sit down." He pointed with his pipe stem to the hard wooden chair before his desk. "And I'm sorry about Laura."

"Thank you, sir."

On the desk in front of him were typed papers, with diagrams annotated with ink in a tight, spiderly crawl I could not fail to recognize.

"Might as well burn them." I pointed at the schematics. "They're worthless."

His gray eyes appraised me. "Obviously, you did not reveal that at the time. Why did you bring him back?"

I closed my eyes.

Laura . . .

In my dreams, she always looked back at me.

"Perhaps it's too soon for you to—"

Opening my eyes, "I beg your pardon, sir." I needed to explain this. "It's my fault. *I knew too much.* I could have solved their problems, with what Feynman had shown me. I could have made the Nazi program work. Put it back on track. Wilhelm would have taken it from there."

"Surely you, of all people, would never—"

"I could, and I would have, sir."

The Old Man was shaking his head, showing too much faith in this poor operative.

I stood up, crossed to the window, and stared outside with hands clasped behind my back, staring at the black statue in the center of the road. Whitehall was a boulevard, I realized, which reminded me too much of Berlin before the war.

"The sporemist infection," I said softly, "is reversible, if caught in time. They would have offered me Laura, and I would have given them everything."

An avuncular hand descended on my shoulder: the first time the Old Man had touched me since we had shaken hands on our first meeting three years and a lifetime before.

"It was why he wanted to come over. Wilhelm, I mean." I sighed. "Too afraid of his masters to remain in place. He'd probably been exaggerating reports of his progress. But if he'd offered them me—"

No need to point out the obvious.

"So our new Nazi friend is worthless," murmured the Old Man.

The smell of wood polish from the paneling; the redolence of pipe tobacco; the cry of a lone seagull gliding outside. None of it made sense.

"You did the right thing," he said.

* * *

Wilhelm Two would never make it to the promised house in Oxford. Instead, out walking on the Marlborough Downs, accompanied by his SOE bodyguards, he would lose his footing on a wet grassy slope and tragically break his neck.

In Whitehall, nobody mourned.

Of course we won the war. The hyperdragons would have made a difference, but for all Wilhelm One's successes, his masters failed to mobilize the resources that would have swung the balance in their favor.

In the closing days, Stalin's forces suffered inordinately heavy losses in their bid to be first into Berlin. To many tactically minded observers, it seemed madness, since the Allied powers had agreed to divide the conquered city between them. But to me, it was obvious that the real target was the institute where Wilhelm Two had worked. Stalin wanted nucleic technology for himself: the weaponry that would later prove its power when the Americans' bomb destroyed so much of the Japanese ecosystem. I could have told Stalin that his men died in vain, but perhaps his own project had been further behind, and whatever knowledge his scientists gained might have been worth the price, in his eyes.

Afterwards, as the joy of V-E Day swept through London's thronged streets, it seemed almost impertinent to think of precious individuals lost during the years of darkness, when so many millions, tens of millions, died.

But that was the point: of the Holocaust victims and the Allied soldiers and the civilians caught in bombing raids, every one had been a real life, defined a world of his or her own.

I tried to hand in my resignation, but where else would I go? To paint, in New Mexico? In that dream, there had been Laura to share my life. Alone, it was cold and pointless.

And so I stayed.

SOE slid into historical oblivion, reinventing itself in peacetime as DI6, then MI6. Even when the Old Man retired, I stayed on throughout the fifties, until I could stand the memories no more.

Every night, I dreamed of Laura.

One day, early in 1962, I resigned.

The next morning—in the intelligence community, no one works out their notice—I sat on a bench in Saint James's Park, reading a *Daily Sketch* that someone had discarded, drinking in the scents of rhododendron I had been too busy to smell when I ran my five miles at dawn.

The newspaper headlines were large. After the western world's paranoia over orbiting Sputniks, this was the sentiment: TIME TO GET OUR OWN BACK.

By "our" own, they meant the U.S.

Journalists focus on personalities, and the man in the spotlight was going to ride a fiery dragon up through the sky, and into the darkness beyond. But, behind the headlines . . . Well, since that notable absence from the Nuremberg trials, I had always known how things would play out.

It's been two decades.

I crumpled up the paper and tossed it into a bin.

Trafalgar Square, when I walked through it, was scarcely changed from the day the Swastika Bomb dropped and Laura's life collided with mine, altering everything.

Whitehall, too, was the same. But, by the time I had passed through Parliament Square and was strolling down Victoria High Street, the buildings were transparent greens and blues and reds, slowly morphing as they cycled through their jelly-forms, while beneath them brightly dressed Londoners rode colorful trexes along boulevards which glistened like a dragonfly's wings.

I remembered Leo's friend Jack, the priest, talking about the importance of history's turning points, and wondered if Mendel's legacy was only now coming into its own, truly defining the shape of modern times.

Westminster Cathedral, like Westminster Abbey at the road's far end, remained in its original form, surrounding by marvels of shifting bioarchitecture, while one- and two-man albatross-glides circled overhead, enjoying the view.

I went up the steps, moved quietly inside.

The high vaulted ceilings were as blackened as before the war, but candles shone brightly and the stained glass windows were magnificent in their rich-hued artistry. Incense was heavy upon the air, as the priest at the high altar celebrated the Mass. Not Jack: he was in South America, on missionary work.

"—art in Heaven, hallowed be thy—"

I turned away, retraced my steps to the entrance.

There was nothing for me here.

I used Thomas Cook's, the real travel agency, to make the booking. The SOE department was long defunct; few people in MI6 now would recognize the term.

Somehow, that made it seem even more appropriate.

Sweltering heat, though it was not yet spring. Blazing sun in a cloudless azure sky. Hotter than New Mexico, and humid.

I lay on straggly parched grass which struggled to grow through sandy ground. Using a soft cloth, I wiped the rifle's scope before putting my eye to the lens once more. A jumble of people: military uniforms, civilian suits. Women in candy-colored frocks, wide hats, and short white gloves. No sign of the target.

Too soon.

I looked away, lay the rifle down, and lay back on the grass. No blimps or scanbats overhead: they were steering clear of the area. The one security hole I was able to exploit.

Florida, and hot: it was over three years since Felix and I had fished here last—he took a giant tuna despite his hook-hand—and I had avoided all contact with him since then. But through him I had made interesting contacts, which was how I was lying here now with an airclaw-loaded rifle, waiting for the moment.

Laura, my sweetest love . . .

Later. Concentrate now.

Huge and magnificent, the great silver hyperasaur stood waiting to lift: vastly bigger than any dragon-flyer built before, pregnant with massive energies, poised to burst upward into sky, into space, to orbit our small planet with its brave lone pilot aboard.

His courage, I saluted.

Nearly time. Adjusting the scope . . .

There.

Among the dignitaries, on the official view-platform overlooking Cape Canaveral, was the one I was looking for. Dressed in a white linen suit, with the same white-blond hair—he had never bothered to have his appearance altered after the war—he took his place beside a five-star general.

The general clapped him on the back and handed over a cigar.

Riches, fame, and the chance to lead the free world into space. Ignoring his past: claiming that his wartime work was carried out in virtual captivity.

A blare of distant tannoys.

Countdown.

Distant roar, as the great hyperdragon's engines burst into life, poured white flames downward, blazing bright as the sun.

Crosshairs . . .

Ready to lift, mankind's future rising to the stars, but my focus now was on a single white-blond head.

Steady.

No Piotr to help me this time: the responsibility was mine alone.

Breathe in . . .

Liftoff.

Even from here, the roar was deafening. Onlookers' cheers lost amid the hyperdragon's thunderous rising.

And hold.

Centered.

It was, I admit it, a triumph for humankind. The first step in our species' fundamental destiny.

Hold . . .

Focus: face, hair, blond-white.

No mistaking the features.

And squeeze.

For Laura.

Turning away, as the speeding silver dragon arced higher, higher on its tail of fire, diminished in the azure sky, was gone.

Acid bubbling: self-destruct dissolved the rifle.

It's over now, my love.

I turned upon the sand-choked grass and walked away.

John Grant is the author of about sixty books (some under other names), both fiction and nonfiction, among them *The Encyclopedia of Fantasy* (with John Clute). His most recent major nonfiction books are *Masters of Animation, Perceptualistics: Art by Jael,* and, with Elizabeth Humphrey, *The Chesley Awards: A Retrospective*. His novel *The Far-Enough Window,* "a fairy tale for grown-ups of all ages," was published in fall 2002, as was his book-length words-and-pictures collaboration with Bob Eggleton, *Dragonhenge*. He has received the Hugo, the World Fantasy Award, the *Locus* Award, the Mythopoeic Society Scholarship Award, the J. Lloyd Eaton Award, and a rare British Science Fiction Association Special Award. Under another name he is Commissioning Editor of Paper Tiger, the world's leading publisher of fantasy art books, and U.S. Reviews Editor of *Infinity Plus*; for his work with Paper Tiger, he received a 2002 Chesley Award. He is married to Pamela D. Scoville, Director of the Animation Art Guild.

NO SOLACE FOR THE SOUL IN DIGITOPIA

John Grant

My wife Xanthe was standing naked at the counter in the bathroom, brushing her teeth, when I woke up. Still lying in bed, I watched her back with a voyeuristic frisson for a few moments; then I climbed out quietly and padded across the bedroom to creep up behind her.

Standing there, looking at her smiling eyes over her shoulder in the mirror, I ran the tips of my fingers softly down over her shoulder blades and then the length of her spine until I reached the lower curves of her buttocks. With my fingernails I repeatedly stroked gently outward from the fork of her legs along the twin creases beneath the smooth swells, all the while brushing her back with the soft hairs of my chest and nuzzling my nose in among the sleepy tousles of coppery red that fell over the back of her neck.

She gave a sort of bubbling moan through the froth of toothpaste and relaxed her stance, moving her feet apart and bending at the waist, supporting herself with her left hand flat on the counter. I stopped my stroking and reached round her hips to let my fingers play among the tangles of her pubic hair, teasing the strands tenderly, then slid one forefinger down to the topmost fold of her sex.

Xanthe's a tall woman, taller than I am, so when my erection slipped between her thighs to find a harbor, it was only the ripe bell of my penis that lodged between her labia, cupped in moist warmth. She carried on brushing her teeth, though her steady rhythm was by now becoming disjointed; stuttering vibrations caressed the head of my penis and transmitted themselves softly down its shaft. The movements of my hands, too, were becoming uncoordinated as I alternately cradled her buttocks and rubbed the rounded ridges of her pelvic cradle, leaning against her once more as I kneaded the sides of her small breasts with the soft skin of my inner wrists, touching the taut raspberries of her nipples and gripping them fleetingly between paired fingers.

She dropped the toothbrush into the basin and gulped, the gulp causing a muscular contraction that tweaked the top of my penis with a little kissing sound. Almost immediately afterwards, further delicate movements there told me that she was having a small orgasm.

She put her other hand down on the counter and, with a long low gasp, bent her legs, easing outward and downward and backward so that the full length of my shaft was slowly engulfed by her hotness and she was almost sitting in my lap as I took part of her weight.

"Good morning," I said in her ear, looking at our faces smiling side by side in the mirror, casting my gaze down a little to where my hands had settled on her breasts. Her skin was pinkly flushed down one side of her neck and over her chest, the pale tint spreading across the top of one breast.

"'Morning," she sighed.

She began to sway herself against me, up and down along my penis, also giving a small swivel to her hips. In the mirror her lips were still forming a contented smile and her eyes were still dancing, but through my own closing eyes I could see my mouth was pulling into an earnest grimace. I dipped my head and began to cover her shoulders with my kisses, my own hips beginning to move now in counterpoint to hers, feeling her smooth buttocks squeezing down onto my lower belly and then pulling themselves momentarily away again. Grunting a little, I ran a hand back down to her sex and found her clitoris with my fingers; the first touch drew a whooping cry from her, and the speed of her movements against me quickened. She shoved backwards, taking a small, uncontrolled step, making me nearly stumble, dropping her head so that now she was almost bent double, pushing me deeper and deeper and more and more forcefully into her, the wet sounds of us slapping the walls with muted echoes.

It was as if there were no longer two of us, just one single organism trembling and pulsing in a quest for unitary pleasure. I felt as if I were being absorbed entirely into her, so that her breath was my breath, her velvety morning skin-smell a warm cloud that embraced both of us, her twitching shoulder blades in front of my eyes a plain of flesh that was my own flesh.

Each plunge of her buttocks against me was now almost like a punch. With each thrust she was yelling—incoherent sounds, wordless cries.

"Shall? We?" I could hardly keep my mind together enough to form the words.

"Do? This?" Someone else might not have recognized her noises as speech.

"Together?" I gasped.

An old joke between us. An affectionate little playful joke it had been at first, in the early days of our lovemaking, but now become so intimate that it was as erotically charged as anything else we did together.

"Yes!" she shrieked.

And began to come, this time not with one of the many little minor orgasms that had been tickling my penis all along but with full intensity, so that it felt as if firm hands were clutching at my shaft. Inside me a dam held momentarily against the rush of my own orgasm, then was breached by a final powerful thrust.

For long seconds my entire universe was just a haze outside the brilliantly glowing focus of our conjoined sexes. I was the smell of her and the taste of her and the touch of her and the breath of her. Then slowly I became conscious that my cheek was nestling against her spine and that I was moaning softly. She was still coming, her orgasms tumbling one on top of each other, my penis lovingly buffeted by their strength, her buttocks moving in spasmic little jerks.The air was rich with the smell of our lovemaking.

Finally she stilled.

I held on to her with both arms around her belly, dizzy, needing her support. I was still hard within her, hard as stone, so hard I thought I would never, ever stop being hard like this, would never stop being in this embrace with her, would be making love with her for eternity and all the while be growing harder. . . .

"I love you," I said.

"I love you," she replied.

Eternity is quite a lot shorter than you think it is, and so half an hour later I was showered and fully dressed and out on the street, walking to the bus stop where I could if I wanted to catch the number 264 bus to work. My whole body felt satisfied, as if it had been given an extensive internal massage complete with sweet-scented oils. I kissed Xanthe good-bye when she stopped at the car, and watched her, waving loosely, as she drove away to join the streams of traffic toward the other side of town, where she worked as a marine architect.

Standing there in the warm morning sunshine, I was somewhat vague as to what precisely my own profession was, and so instead of carrying on toward the bus stop I turned up a small alley and was in bed with my wife Lyssa.

Lyssa is small and blond and so petitely cute that it almost hurts the eyes to look at her, as if so much beauty shouldn't have been crammed into such a small compass, and right at the moment she was coming drowsily awake with her face burrowing into my side. I pushed silky near-white hair back from her forehead and squirmed around so that I was facing her and could kiss her into full wakefulness. Even as her eyelids pulled slowly open, her eager little hand was tickling its way down over my belly to where my erection was slowly, then swiftly, mounting.

She wrapped her hand around the shaft, pressing her thumb down on top of the tip, grinning through the increasingly fervent kisses I was placing on her lips. Gripping a little more tightly, she began moving the thick satiny skin of the shaft slowly up and down.

"The kids are awake," I whispered, pulling my face back from hers. "I heard them moving around a little while ago."

She pouted, half-mocking, half-serious. "Damn," she said.

When Lyssa and I make love, it tends to get noisy. The kids know better than to charge into our bedroom unannounced, of course, but they'd be bound to ask us later about all the creaking and yelling. We've tried making love silently, and it's kind of fun because it feels as if we're committing some sort of delicious naughtiness together; but at the same time inhibited lovemaking has its limitations.

"Still and all . . . ," I began, leaving the sentence unfinished as I rolled her onto her back.

I kissed her again, harder this time, our tongues probing into each other's mouths, caressing within, our breath becoming loud in each other's ears. Then I ran the tip of my tongue around the edge of her cheekbone and down to her chin. She arched her head back as my tongue moved on to the crease of her neck, tasting her sweetly salty sleep-sweat there. In moments I was kissing her breasts, taking each of her small nipples in turn between my teeth, not biting, not even gently biting, rolling my lips over my teeth and pressing gently. She held my head in her hands, breathing more deeply now, her hips moving reflexively against my chest, her legs parting to either side of my torso, her furry little sex making small wet noises of its own against my abdomen. Between her legs, my erection against the sheets seemed heavier than it had any right to be, as if it were becoming larger than all the rest of me put together.

I lingered at her breasts tantalizingly longer than I knew she wanted me to; then my mouth continued its downward exploration of her body, pausing once more when I reached her navel, probing the tiny folds with my tongue, tracing damp circles around the concavity. Her belly is flat and smooth, even though she has borne our two children; through its wall I could hear her blood rushing and her insides making modest gurgling noises. Often enough in the past I'd held my cheek here and listened to first my son and then my daughter moving in her womb, felt their diminutive kicks and punches. I knew I was smiling broadly as I kissed her here now, sensing the tension building up in her loins, remembering those other moments of intimacy. . . .

Half-kneeling, I pushed her legs gently yet farther apart and sank my head to rub my nose over the silky skin of the inside of first one thigh and then the other, breathing deeply the warm muskiness of her pinkly unfolding sex. Her fingers began clawing the sides of my head as she tried to force my face closer, but I deliberately resisted her just a little while longer, continuing to rub and kiss her inner thighs as her behind squirmed frenziedly against the sheets.

At last I relented, first placing a broad warm kiss over her wet opening, sucking slightly, tasting the juices; then I puckered my lips and very daintily kissed her clitoris. She jerked against me, even that brief and inconsequential contact bringing her to the brink of orgasm. Once more I kissed the little nub of her clitoris, and then I pushed my tongue between her warm folds, blowing gently.

Her hands scrabbled over my scalp. Her hips jolted upward, clear of the bedding, and despite any resolves to keep silent she let out a long, soft cry as she came, her thighs juddering against my ears, the muscles of her belly writhing and tautening.

When she was at last done, I pulled myself up alongside her, my penis, though somewhat smaller here than usual, still a dominating presence in my mind. I gazed at her reddened face, at the beloved crinkles of her ears, at those pale blue eyes still half-closed, at the parted pink lips, at the corner of her mouth where a trickle of saliva glistened, at the darling curls of damp-darkened blondness flattened against her scalp around the back of her ears. I truly love this woman of mine—love every part of her, both spiritually and carnally. I caressed her cheek and jaw with my hand, feeling her heat and her perspiration, and, despite the throbbing of my erection, began to drift into a half-doze, cradling her in my arms against me.

But then her hand walked its way down to my erection again and clasped it.

"Wait here," she breathed.

Pushing back her hair with her other hand, she wormed her way slowly down the bed and kissed the underside of my penis, working slowly up from root to tip, where she tickled with her tongue tip the tight cord of folded flesh.

I shut my eyes and saw visions of rushing through space among the stars, my consciousness ebbing toward the tingle of her tongue tip's touch and the warmth of her breath.

The movements of her tongue grew broader, and then she was taking me into her mouth, almost coyly at first, as if her mouth were experimenting demurely; then more fully, her lips gripping the skin and sliding it backward and forward, a finger and thumb gripping the base of the shaft to steady it.

Within just a few seconds I felt myself beginning to come. Lyssa felt it as well, starting to press rhythmically with her thumb at the throbbing area near the base of my erection's underface. She speeded up the motion of her head, clamping my penis more firmly between her lips, her tongue dashing hither and thither.

I craned my neck and looked down along the planes of my body at her. Just at that moment she glanced up at me, paused for a fraction of a second, her blue eyes alight with mischief, her cheeks puffed out. Her lips moved in a languorous grin, and I felt every last adjustment of her flesh against mine. And then I flung my head backwards on the pillow, arching my back, staring at the gray ceiling, groping with my hands for her head, for her hair, running my fingers through it, feeling the play of her facial muscles and the subtler tectonics of her scalp, putting a palm to her bulging cheek and . . .

The breath rushed out of my lungs in a huge gust as I came, feeling her swallow once, twice, her head now stilled as she accepted my semen.

Once she had satisfied herself that she'd drained me entirely, she fastidiously licked me clean, touching me gently with her lips and her tongue because she knew how close to pain any rough contact would be. Then, holding her body against mine, she slithered up the front of me until our mouths were together and I could kiss her deeply, tasting myself on her teeth and tongue.

"I love you," I said at last.

"No more than I love you," she said, her voice hoarse, hardly more than a whisper. "Hold me tight, darling. Hold me tight."

I held her tight, pulling a sheet up over us in case the kids came bouncing in.

It was a Saturday, so the four of us breakfasted together in leisurely fashion as the sunlight spilled in yellow pools across the kitchen. Lyssa scrambled some eggs—daughter Karen's favorite—while I prepared endless relays of hot toast for Mark and myself. Every now and then Lyssa and I would catch each other's gaze above the industrious small heads of the scarfing kids and share a secret grin with each other. After a while I had to sit down at the table with the kids because those intimate grins of hers were making my weary, still slightly throbbing penis swell again inside my slacks. I diverted my mind by making earnest conversation with Mark about the iniquities of various of his schoolteachers.

After breakfast was over, the table and counter wiped clear of crumbs and smears, the dishes stashed safely away in the dishwasher and the kids safely out at play in the yard, Lyssa and I clung together for a long moment, feeling the familiar curves and angles of each other's bodies through the resented barriers of our clothing. If it hadn't been for the kids, we'd have made love once more, there and then, on the floor or the kitchen table. My fingers on the crotch of Lyssa's pale blue jeans, I could sense her renewed dampness. My own excitement was far more obvious.

She touched a hand to my lips.

"Later, later," she said with a lopsided smile. "You've got a date for tonight, mister. Don't forget."

I kissed her one final time and made for the door.

Whether it was the afterglow of love or whether it was carelessness, or whether it was simply an unexpected quantum event—as if that weren't a tautology—I do not know, but as I walked down the bright dusty street of our semirural small town, my mind anticipating eagerly a visit with my wife Isolde, my feet took a wrong turn and I was sitting in a cluttered one-room apartment listening to poundingly loud music that I eventually recognized as Bryan Adams.

Damn!

These things sometimes happen—in fact, probably more often than anyone quite likes to admit. The pathways through the interstices of the polycosmos are more than infinite in number, so it's hardly surprising that our instinct, being powered by neurons that are somewhat, if only slightly, less in number than those pathways, should occasionally lead us astray as we journey between the realities. Usually it doesn't matter at all, of course: you emerge to discover yourself with a partner of either sex who is, shall we say, similarly inclined, and whom you have always and forever known and loved. Bathing in the mutual love, either alone together in bed or on a long country walk with the dogs or yelling and screaming with the kids on a carnival roller-coaster—or wherever, all the myriad ways there are in which love can be made—you establish your presence, your soul, your identity, so that the whole great machine of the polycosmos keeps trundling along toward its eternal futures.

But sometimes you're not so fortunate with these missteps of the mind, and this was what had happened to me now.

I had landed myself in one of those realities I call digitopias.

If you're lucky or simply more stay-at-home than most of us, you've never encountered a digitopia. They're rare—*thankfully* rare. So let me explain.

Everybody knows that the totality of existence comprises the polycosmos, which is the sum of all the infinity of infinities of realities that there are or ever could be. Each of those individual realities can be considered as not just a universe of its own but also a microcosmic (if you can sensibly use such a term of entities that are so vast) polycosmos in itself, for it, too, comprises countless infinities of realities. And, of course, each of those realities is in itself a further polycosmos. . . .

Every event, every action, every slightest shift of a quark will—and does—spark a new reality, yet there are also infinitely many and infinitely large *families* of realities brought into existence at the birth-moment of each of the universes; these families are like the major branches of some unimaginably huge and complex tree, with the other realities serving as its twigs and buds and leaves. What distinguishes one family from the next is that the physical laws governing each are not quite the same as in any other family. At the birth-moment, the Big Bang, of each universe/polycosmos-in-the-making there springs into reality whole gamuts of possible sets of physical laws under the reign of which a viable universe might run, and *all* of those viable sets are reified.

The most important physical law that can be varied from one to the other concerns the rate of passage of time: as a second passes in one, a year might pass in the next, or a millennium. If you cared to, you could pick a pathway through the interstices between the realities of even just our own small universe so as to travel into realities where everything is many billions of times older; or, by going "the other way," you might enter realities (if you were fool enough to do so) where the Big Bang initiated itself scant milliseconds before. It's as if you could travel freely backward and forward through time, except that all the realities and subrealities are different existences—they've evolved independently and, of course, at variant rates, and some have been majorly influenced by trains of events that have never even touched others.

Still, if you move carefully among the interstices, you can restrict your travels to realities—whether within our universe or elsewhere in the polycosmos—where the speed of time's flow is always much the same, and where the physical laws, too, are not so divergent from those you are accustomed to. The sun is still yellow-white in a sky of blue. When you drop a dish, it falls swiftly enough to smash on the floor—or at least it does *fall,* rather than drifting off to the side or floating blithely up and away.

But some of the very tiniest variations in the physical laws can have the most profound consequences, and they can make two realities that should, on the face of it, be nearly identical in fact be strikingly different.

The behavior of the electrons of the antimony atom is one area of relevance. In almost every reality where there exists such an element as antimony at all, which means 99.99999999 percent repeating of the realities within our own local polycosmos, the polycosmos that we'd regard as our universe's families of realities—the electrons squat lumpenly around the nucleus, unwilling to be dragged away from it. In a few realities, however, those electrons are prepared to be far more freewheeling, and in a tiny percentage of *those* miserably unfortunate realities it has been discovered that antimony can be used to "dope" silicon such that minuscule transistors may be manufactured.

The reason I call such godsforsaken realities digitopias is that they have a science of microelectronics—tiny computers everywhere, gadgetry proliferating like a plague of locusts and devouring all the soulstuff of the people of the worlds.

The reason I knew I'd accidentally danced into a digitopia was that the Bryan Adams track that was filling my ears was being played by a small black machine with on its front a display of moving lights that served no purpose.

And so on.

Well, there I was, dumped like a beached whale into a digitopia. I could have turned straight around and left—I could physically have forced myself to do that—but it would have been to violate the rules of love that keep the polycosmos (and the polycosmos of polycosmoses, and all the tiny polycosmoses that swarm within and alongside every particle of us) evolving and growing and alive. I had to stay here for at least long enough to give whatever love I could to the partner who awaited me, and whom I had known and loved for a long, long time.

I was sitting on the floor of the apartment, fully clothed, my back against the wall beneath a curtained window. Opposite me, also sitting with her back against the wall and also fully clothed, was my longtime fuckmate (her term) Kath. I'd asked her several times if maybe, you know, we should get married, but her answer was always the same: she didn't want to commit herself to anything too long-term that she might not be able to sustain. Only once did I try to explain to her that there was a way of having both permanence in love and yet also constant change and freshness.

She'd looked at me as if I were crazy.

"You've seen it yourself," I'd said. "You must have. You know those moments when your gaze meets someone else's across a crowded room or on a train, and you all at once realize that, even though you've never met or spoken to this person and probably never will, you *recognize* them as somebody you've known intimately, body and soul, for as long as you can remember. And you can see them recognizing you, as well, in the same way."

"You fancy them, like?" said Kath, struggling to grasp what I was saying.

"No, it's not that at all. What the pair of you suddenly realize is that you've spent a whole love story together, but not here, not now, not in *this* reality—so it's a love story you'll never be able to read with your lives, a love story that you'll never know."

Kath began to chortle. "You're just saying you sometimes fantasize about fucking other women, breaking it to me gentle, like. But that's OK. Sometimes I fantasize about fucking other guys, so we're equals. Quits. You hungry?"

(I remembered all these things, you see, even though I'd never been here before.)

"No," I insisted. "Sure I sometimes wonder what it'd be like screwing pretty women I see in the street or meet at parties—I'm not a saintly monk, or whatever—but I'm not talking about that."

"What *are* you talking about, then? You keep saying what you're *not* talking about, but what you're saying doesn't make much sense. Not *really*."

"I'm talking about the encounters all of us sometimes have with our lovers, but our lovers in realities other than the one we happen to be in. Maybe we've been able to cross into those different realities briefly in dreams, or something, or maybe it's just that our awareness can seep through the boundaries that separate the realities—I don't know. But it's a fact. Those little instants of recognition—they're a *fact*."

"You've been watching too much *Star Trek*."

And that was as far as the conversation went.

Gazing across at my fuckmate now, I felt my heart lift with song and loving, because she was very, very beautiful. Her hair was a cascade of black curls surrounding her oval face. Her skin was gray with the weariness of existence, and I wished I could do something to change that; but for all that she was lovely.

"Kath," I said as the CD reached its end and the machine grunted and squealed softly, changing to another. "Kath, come here. I want to hold you."

"Nah, not right now," she said. "I wanna spend some time with the veerigogs. Wanna join me?"

I sighed.

Veerigogs. VR goggles. The latest craze in this dismal digitopia. Kath had bought (or, I suspected, stolen) us each a pair as soon as they'd started appearing in the high-street electronics stores. Put on a pair of veerigogs and you enter a fake, or virtual, reality. If you have the right plug-in-and-play program chips, you can choose any fake reality from the large palette of scenarios the manufacturers have devised. You can use a pair of veerigogs on your own and the illusion seen by your eyes and heard through microspeakers in your ears is so all-embracing, so entirely convincing, that your other senses are deceived as well: you can touch the surfaces, you can smell the odors, you can taste the tastes of the imaginary world. But the way most people use veerigogs is together, fucking with each other while their senses tell each of them individually that they're fucking someone else: a movie star, or a rock singer, or . . .

When Kath had teased me about fantasizing fucking other women, she was counting the use of veerigogs as fantasizing. But there's a difference between a fantasy and a fake reality.

"OK. I guess so." My voice sounded eager not because I was skilled at disguising my true lack of interest but because, by definition, I loved this fuckmate of mine and had done so for as far back as my memory would go: giving her pleasure, in whatever way I could, was to gain pleasure for myself.

She tossed me my veerigogs and began to haul herself out of her overalls. I caught the gadget and began myself to undress, staring at her body as it was unceremoniously revealed to me.

There's an old cliché that some people are more beautiful with their clothes on than off. It's not something that I've ever much subscribed to, because I always love my wives wholly, and find them equally beautiful however I might see them. But, as I watched Kath, I was able to distance a part of my mind and see her as someone else might do, and that facet of me saw that she was a woman who was startlingly more beautiful naked than even her lovely, adorable clothed self. Her legs were long—longer even than Xanthe's, and sleeker. Her entire body, had she ever learned to straighten it properly, had a dignity that was somehow chaste while at the same time deeply sensuous. I wanted to spend days just poring over each square inch of her skin, even the places where she'd bruised herself as a result of her habitual clumsiness. I wanted to lay my head in the small of her back and stare at the smooth, perfect twin drifts of her behind rising like hillsides, their foothills just inches from my eyes, the downy hairs on them making scintillating stars as the breeze blew across them in the sunshine. I wanted to try making love with her through simple touching—the two of us stroking each other's faces and flanks, gazing all the while into the depths of each other's eyes and souls, until our sensualities bubbled over. I wanted her to be astride my face so that I could lose myself in the warm world of her sex, kept from floating away into the ether only by distant sensations of her mouth on my own sex. I wanted to hear the loud pulsing music of orgasm, see the wraithed colors it brings, like trails of illuminated dust clouds out where the stars are young. I wanted us to dance together in our lovemaking to the place where there are no longer two, but one.

"Who do you fancy me being this time?" said Kath, grinning. Her breasts were pale little apples, pink-tipped, framed by brassiere creases, but she was unconscious of them as she fiddled with the fastening at the side of her veerigogs. "Marilyn Monroe? Michelle Pfeiffer? Winona Ryder?"

All I want you to be is Kath, I thought, but of course I didn't say the words out loud because to do so would have been to hurt the one I adored.

"Natalie Marahat," I said at random. Marahat had recently starred across from Richard Gere in a frothy little comedy we'd seen at Loew's. She has a grace and style and slow-moving elegance and profound beauty, not to mention the most gorgeously appealing little smile—all of which were completely wasted in that movie. Somewhere in the polycosmos, I'd realized as I'd stared at the big screen amid the popcorn redolence of the movie theater, perhaps she and I had loved or would love each other—in truth, *did* love each other, because there's no past or future in real love, just an everlasting present.

Kath chuckled. "In that case, then, I guess you'd better be Richard Gere." She plugged the cord of her veerigogs into the box and tapped in instructions. "He's a bit old for my normal type but . . . yes"—reading the microscreen—"I can have him when he was still thirty." She beckoned to me to plug my own cord in. "And here's Natalie Marahat for you." All cozy. There, that's settled.

Standing close to her nakedness, as I now was, I'd developed a quite enormous erection. I don't mean just that it was hard as a rock—although it was—but that in this reality I had a bigger penis than in just about any other I'd so far been in. My balls were pretty gargantuan, too, dangling in a sac that seemed to stretch halfway to my knees, but my erection really was a truly impressive object. In fact, although not so impossibly huge that sex wouldn't be practicable at all, it was probably too big and thick to be a very satisfactory part of genuine lovemaking, so I was quite relieved that it belonged in a digitopia rather than be an encumbrance elsewhere.

Kath didn't seem to notice it, concentrating as she was on fine-tuning the physical attributes of Richard Gere as well as setting the controls to define the fake environment we'd find ourselves in.

"Now," she said at last.

We lay down together on our futon and carefully donned our veerigogs, trying to synchronize as nearly as possible our entrance into the artificial world.

I was on a very good imitation water bed. Lying there beside me, naked in the candlelight, was a truly lovely creature, her body flawless and perfectly proportioned. Natalie Marahat.

A fake Natalie Marahat.

What made the fakery obvious was exactly that: the flawlessness, the absolute perfection of her every proportion. It was as if her body had subliminal erotic triggers implanted in every pore. The result, instead of making my mind blaze with desire, was an overload of erotism that brought into stark blatancy the chill of her artificiality.

As this vision of loveliness put her arms around me, I thought as fervently as I could of the naked Kath I had left just moments before.

"Fuck me, baby," said Natalie Marahat.

And so I fucked her.

Later, after we'd taken the veerigogs off, I tried to coax Kath into lovemaking on the futon, but she said she was tired, and anyway *The West Wing* was just about to start on television.

I left her during the first commercial break.

She didn't notice me going, of course, because there was still the familiar slumped male on the couch beside her, nibbling crackers as he watched a woman on the screen tell him how he could say goodbye to all his allergies with only a long list of minimal possible side-effects. I slipped on a pair of jeans and a sweatshirt and a pair of old, worn-down sneakers, and made my way out of the apartment and down three flights of stairs and out into the streets of a city that was a blare of light and sound. I sensed that everywhere around me there were people busily watching other people doing things that they wished they could do themselves, or plugged into one device or another that would effectively obstruct their every fortuitous leaning toward discovering the ways you can skip through the interstices of the polycosmos, pausing wherever you will to fuel it with the love it needs to keep on living and becoming. All of these machines, driven by the properties permitted by the laws of physics to belong to the electron shell of the antimony atom in only a tiny percentage of the realities that make up the everness—all of these machines pretending to stimulate yet instead inhibiting the imagination from roaming through the infinite possibilities that the polycosmos reifies.

I stood on a street corner, listening to the clangor, and offered up my thanks to the surly dark artificial orange sky that digitopias are so infinitesimally few among the boundless realities.

And at the same time I felt a great rush of pity for the individual souls trapped here.

Souls?

Half-souls, quarter-souls, shriveled relics of souls, more like. Their only blessing is that they don't recognize their trappedness, believe instead that their undiscovery of love represents the fullest freedom of them all. They're prisoners who aren't conscious of the cage's bars, who look at the window of their cell and see just reflected a gray light that they believe is the world . . . and see nothing of the riot of colors and brightness beyond the razor-thin imprisoning glass they could shatter with a touch.

I turned away from the scowl of the angrily uplit sky, took a few paces along the sidewalk, and was in a field of bright green grass with my many husbands. I called out to them in merriment, and they grinned and waved at me, then began to walk toward me, singly or severally.

The insides of my thighs were already damp by the time the first one reached me.

I loved them, I loved my husbands, and they loved me, their wife. We always love each other.

And today, with the sun shining down upon us and the blue and yellow flowers bobbing their heads in the breeze that caressed the grass, we had a lot of lovemaking to do if the infinite realities of the polycosmos were to continue eternally to bring forth their blossom.

Pat Cadigan is the two-time winner of the Arthur C. Clarke Award for her novels *Synners* (reissued in 2001 by Four Walls, Eight Windows) and *Fools* (Bantam Spectra 1992; HarperCollins UK 1994). Her next novel is *Reality Used to Be a Friend of Mine* (Macmillan-UK). She is all wired up in North London and never, ever works without a Net. Better living through chemistry! Technology is our friend! And while we're at it, fax your congressperson; save Internet radio!

AFTERWORD: LIVING WITHOUT A NET?!

Pat Cadigan

You've got to be kidding, I said. Write a story set in a world without a Net? But it wasn't a joke. Which is too bad. When it comes to the Net, I can always use a good laugh.

The Net and I are locked in a struggle that won't end until one of us goes down and stays down. And this time, it's personal.

Trying to work out this tempestuous relationship has brought me to the brink of madness more times than the most difficult of my difficult relatives. Oh, for the simplicity of a science fiction universe. In a science fiction universe, when people invent stuff, it by God *works* and everybody just uses it. You don't get malfunctions unless it's absolutely vital to the plot. And even if things don't turn out quite how everyone wanted, you still feel that you've lived through something significant.

Meanwhile, back in real life, there's an upgrade going on that means over 60 percent of Company A's subscribers won't be able to pick up their e-mail for six hours (give or take a day); lightning knocked out the southeast sector of Cable TV Company B, including the broadband Internet service. If you're out shopping or drinking between the hours of 4 and 6 P.M., allow extra time when paying the check by credit card—it's rush hour and the charge-approval lines are so jammed under the influx of calls, it takes twenty minutes just to get a busy signal. Here in my study, right on my very desktop, my computer is suddenly insisting that the modem it's been using all day doesn't exist, never existed, and no one can prove otherwise.

But hey, lest we forget, the technology wasn't *always* this good.

Recently, in my travels around London, my hometown-of-choice, I came upon the following symbols chalked on the side of a wall:

)(

Having spent the better part of the morning answering my e-mail, I thought at first that emoticons were stalking me off-line. However, the plate-o-shrimp law* was in effect; a few hours later, I was reading a news item about something called *warchalking,* which traces its lineage back to symbols used by hoboes to alert fellow travelers as to the hospitality and hazards of the immediate locale. But these symbols are for high-tech hoboes with cleverly designed hardware, functioning software, and instincts that tell them something's in the air.

In other words, there are little pockets of urban territory where you can hitch a ride onto the Web courtesy of someone else's network. Look, Ma, no wires.

Well, you can if you know how, and what to look for. So if you know how, and you see this,

you've found an open node.

*Discovered ca. 1983 by the weirdest character in *Repo Man,* which states that at any given moment, an apparently random thought entering a mind in the waking state will cause its three-dimensional real-world counterpart to manifest regardless of circumstances or location. This manifestation usually occurs within a time frame of anywhere between (roughly) thirty minutes and 120 minutes—although time periods as short as thirty seconds and as long as twenty-four hours have been recorded in cases of extreme hunger (the former, thus, "plate-o-shrimp") and long-lost relatives/associates/nemeses (the latter, e.g., "I was just thinking about you . . .").

This is a very exciting development, I hear, which will no doubt help usher in the Golden Age of Wireless, twenty-first-century version. Talk to some of the more technoculturally tuned-in Big Brains (many of whom wrote stories for this anthology), and I'm sure they'll be able to rattle off a long list of wonders that could come out of this, and I'd be tempted to bet money on a lot of them coming true.

But I know what else is going to happen: drive-by spamming.

The image I can't get out of my mind is that of reading the news on my metallic red super-duper-PDA (I'll get one someday) when suddenly the high-res color screen starts lighting up with offers for everything from surplus aluminum siding to Viagra in bulk. Along with, of course, come-ons for software guaranteed to lock out any drive-by spam as well as software guaranteed to confound any drive-by spam lockout.

And when I press DELETE, my super-duper-PDA has two seconds to inform me there isn't enough memory to carry out that command before the screen catches fire.

Perhaps this comes out of a personal failing on my part. If I were a stronger person, all the bugs, the mysterious error messages, the page faults and the stack overflows, the crashes, the blue screens of death, the black screens of no-kidding death, and the uncounted multitudes of pages that go *not found* without warning or explanation (there must be millions of them by now) wouldn't get to me like this. I'd just laugh them off. Well, I'm trying. But the Net is a lot more trying.

At this point, some techie somewhere is protesting that I'm indiscriminately mixing hardware and software issues with Net issues. Listen, I'm not confused or disorganized—I'm holistic. If you want all the parts to work together, you have to work with them as a whole unit. I mean, really—I suffered from math anxiety compounded by childhood exposure to the new math, but even *I* know you don't do anything to one side of an equation without doing the same thing to the other side.

All right, I know what you're thinking: *We can see that you have issues, but what's the Net got to do with it?*

It's like this: I've imagined some pretty wild things in my twenty-five years as a professional SF writer, and not all of it is directly attributable to things I did in my misspent youth. But when I was approached to come up with a story for a world without a Net, I had nothing to say.

I've spent decades imagining a world *with* a Net. I mean, a Net that *works*. I mean a Net that *really* works, the way useful tools are supposed to work. Yes, I know: it's not a perfect world. That's the truth—therefore, it isn't even fiction, let alone science fiction.

Yes, I know—that's not the only kind of science fiction there is. In fact, you're holding a whole book of alternatives. But no matter how I looked at it, I kept coming to the same conclusion:

A world without a Net? Been there. Still there.

Could somebody please e-mail me when it's working?

Lou Anders is an editor, author, and journalist. In 2000, he served as the Executive Editor of Bookface.com, an Internet company that provided books and short stories for free online reading, and before that he worked as the Los Angeles liaison to Titan Publishing Group. He has published more than 500 articles in such magazines as *International Studio*, *Dreamwatch*, *Star Trek Monthly*, *Star Wars Monthly*, *Babylon 5 Magazine*, *Sci Fi Universe*, *Doctor Who Magazine*, and *Magna Max*. He is the author of *The Making of Star Trek: First Contact* (Titan Books, 1996), and the editor of the anthology *Outside the Box* (Wildside Press, 2001).